Correction Techniques
in Emission Tomography

Series in Medical Physics and Biomedical Engineering

Series Editors: John G Webster, Slavik Tabakov, Kwan-Hoong Ng

Other recent books in the series:

Practical Biomedical Signal Analysis Using MATLAB®
K J Blinowska and J Żygierewicz

Physics for Diagnostic Radiology, Third Edition
P P Dendy and B Heaton (Eds)

Nuclear Medicine Physics
J J Pedroso de Lima (Ed)

Handbook of Photonics for Biomedical Science
Valery V Tuchin (Ed)

Handbook of Anatomical Models for Radiation Dosimetry
Xie George Xu and Keith F Eckerman (Eds)

Fundamentals of MRI: An Interactive Learning Approach
Elizabeth Berry and Andrew J Bulpitt

Handbook of Optical Sensing of Glucose in Biological Fluids and Tissues
Valery V Tuchin (Ed)

Intelligent and Adaptive Systems in Medicine
Oliver C L Haas and Keith J Burnham

A Introduction to Radiation Protection in Medicine
Jamie V Trapp and Tomas Kron (Eds)

A Practical Approach to Medical Image Processing
Elizabeth Berry

Biomolecular Action of Ionizing Radiation
Shirley Lehnert

An Introduction to Rehabilitation Engineering
R A Cooper, H Ohnabe, and D A Hobson

The Physics of Modern Brachytherapy for Oncology
D Baltas, N Zamboglou, and L Sakelliou

Electrical Impedance Tomography
D Holder (Ed)

Series in Medical Physics and Biomedical Engineering

Correction Techniques in Emission Tomography

Edited by

Mohammad Dawood
European Institute of Molecular Imaging
Münster, Germany

Xiaoyi Jiang
European Institute of Molecular Imaging
Münster, Germany

Klaus Schäfers
European Institute of Molecular Imaging
Münster, Germany

CRC Press
Taylor & Francis Group
Boca Raton London New York

CRC Press is an imprint of the
Taylor & Francis Group, an **informa** business
A TAYLOR & FRANCIS BOOK

CRC Press
Taylor & Francis Group
6000 Broken Sound Parkway NW, Suite 300
Boca Raton, FL 33487-2742

First issued in paperback 2019

ISBN-13: 978-1-4398-1298-3 (hbk)
ISBN-13: 978-0-367-38144-8 (pbk)

Library of Congress Cataloging-in-Publication Data

Correction techniques in emission tomography / [edited by] Mohammad Dawood,
    Xiaoyi Jiang, Klaus Schäfers.
        p. cm. -- (Series in medical physics and biomedical engineering)
    Includes bibliographical references and index.
    ISBN 978-1-4398-1298-3 (hardback)
        1. Tomography, Emission. 2. Radiative corrections. I. Dawood, Mohammad. II. Jiang,
    Xiaoyi. III. Schäfers, Klaus.

    RC78.7.T62C67 2012
    616.07'57--dc23   2011049866

Visit the Taylor & Francis Web site at
http://www.taylorandfrancis.com

and the CRC Press Web site at
http://www.crcpress.com

Contents

About the Series   xi

Foreword   xv

List of Contributors   xix

1 Introduction   1
   Klaus P. Schäfers
   1.1 Introduction   1
   1.2 Principle of emission tomography   2
   1.3 Electromagnetic spectrum   4
   1.4 Need for correction techniques   4
   References   7

I Background   9

2 Biomedical Applications of Emission Tomography   11
   Michael Schäfers, Sven Hermann, Sonja Schäfers, Thomas Viel,
   Marilyn Law, and Andreas H. Jacobs
   2.1 The role of imaging in biomedical research
         and applications   11
   2.2 Functional and molecular imaging by emission tomography
         enables high sensitivity and spatial resolution 13
   2.3 Biomedical applications of emission tomography depend
         on tracers   14
   2.4 Applications   16
         2.4.1 Preclinical applications   16
         2.4.2 Clinical applications   17
         2.4.3 Examples of biomedical applications of emission
                 tomography   18
                 2.4.3.1 Bioluminescence imaging of tumor growth . .   18
                 2.4.3.2 Dynamic PET in pharmakodynamic studies .   19
                 2.4.3.3 From mice to men—Non-invasive translational
                           imaging of inflammatory activity in graft-
                           versus-host disease 20

           2.4.3.4   PET to quantify catecholamine recycling and
                     receptor density in patients with arrhythmias   22
           2.4.3.5   Multiparametric imaging of brain tumors . .   23
    References   26

3 PET Image Reconstruction   **31**
    Frank Wübbeling
    3.1 Introduction   31
    3.2 Analytical algorithms   32
         3.2.1 Mathematical basis   32
         3.2.2 Filtered backprojection   35
         3.2.3 Implementation: Resolution and complexity   37
         3.2.4 Implementation and rebinning   38
                3.2.4.1   2D Rebinning   39
                3.2.4.2   3D filtered backprojection   40
         3.2.5 Limitations   40
    3.3 Discrete algorithms   40
         3.3.1 ART—Algebraic reconstruction technique   41
         3.3.2 EM   42
         3.3.3 Computing the system matrix   44
         3.3.4 List mode   45
    3.4 Summary   47
    References   47

II Correction Techniques in PET and SPECT   **49**

4 Basics of PET and SPECT Imaging   **51**
    Ralph A. Bundschuh and Sibylle I. Ziegler
    4.1 Introduction   51
         4.1.1 Interaction of photons with matter   52
                4.1.1.1   Photoelectric effect   52
                4.1.1.2   Compton scattering   52
         4.1.2 Photon attenuation   54
         4.1.3 Scatter 57
         4.1.4 Variation in detector efficiency, normalization . . .   58
         4.1.5 Dead time effects (loss of count rate)
                (PET and SPECT)   59
         4.1.6 Partial volume effects (PET and SPECT)   59
                4.1.6.1   Spill out   60
                4.1.6.2   Spill in 60
         4.1.7 Time resolution and randoms (PET only)   61
         4.1.8 Collimator effects—Distance dependent spatial
                resolution (SPECT only)   62
         4.1.9 Positron range and annihilation (PET only)   63
    References   64

5 Corrections for Physical Factors   **67**

Florian Büther

5.1 Introduction 67
5.2 Decay correction 69
5.3 Randoms correction 71
     5.3.1 Singles-based correction 72
     5.3.2 Delayed window correction 72
5.4 Attenuation correction 73
     5.4.1 Stand-alone emission tomography systems 77
     5.4.2 PET/CT and SPECT/CT systems 80
     5.4.3 Attenuation correction artifacts 82
5.5 Scatter correction 90
     5.5.1 Energy windowing methods 91
     5.5.2 Analytical methods 92
     5.5.3 Direct calculation methods 94
     5.5.4 Iterative reconstruction methods 95
5.6 Concluding remarks 95
References 95

6 Corrections for Scanner-Related Factors   **105**

Marc Huismann

6.1 Positron emission tomography 105
     6.1.1 Introduction 105
     6.1.2 Data normalization 107
     6.1.3 Noise equivalent count rates 108
     6.1.4 System dead time 108
     6.1.5 Partial volume 110
6.2 Single photon emission computed tomography 112
     6.2.1 Linearity, center of rotation, and whole body imaging 112
     6.2.2 Motion correction 114
References 115

7 Image Processing Techniques in Emission Tomography   **119**

Fabian Gigengack, Michael Fieseler, Daniel Tenbrinck, and Xiaoyi Jiang

7.1 Introduction 119
7.2 Denoising 121
     7.2.1 Image domain 122
     7.2.2 Fourier transform domain 123
     7.2.3 Wavelet transform domain 124
7.3 Interpolation 126
7.4 Registration 129
     7.4.1 Categorization 130
         7.4.1.1 Nature of transformation 132
         7.4.1.2 Similarity measure 133
     7.4.2 Validation 135

   7.4.3 Software 137
  7.5 Partial volume correction 137
   7.5.1 The partial volume effect in PET imaging 138
   7.5.2 Correction methods 140
  7.6 Super-resolution 144
  7.7 Validation 146
   7.7.1 Intensity-based measures 146
   7.7.2 Phantoms 148
    7.7.2.1 Hardware 148
    7.7.2.2 Software 149
  References 150

8 Motion Correction in Emission Tomography   157
  Mohammad Dawood
  8.1 Introduction 157
   8.1.1 Magnitude of motion 158
    8.1.1.1 Patient motion 158
    8.1.1.2 Respiratory motion 158
    8.1.1.3 Cardiac motion 159
  8.2 Motion correction on 3D PET data 160
   8.2.1 Overview 161
   8.2.2 Rigid motion correction 162
   8.2.3 Elastic motion correction 163
  8.3 Optical flow 164
   8.3.1 Image constraint equation 164
   8.3.2 Optical flow methods 166
   8.3.3 Optical flow in medical imaging 167
  8.4 Lucas–Kanade optical flow 168
  8.5 Horn–Schunck optical flow 169
  8.6 Bruhn optical flow 170
  8.7 Preserving discontinuities 172
  8.8 Correcting for motion 173
  8.9 Mass conservation–based optical flow 174
   8.9.1 Correcting for motion 175
  References 177

9 Combined Correction and Reconstruction Methods 185
  Martin Benning, Thomas Kösters, and Frederic Lamare
  9.1 Introduction 186
  9.2 Parameter identification 187
   9.2.1 Compartment modeling 187
   9.2.2 4D methods incorporating linear parameter
     identification 189
   9.2.3 4D methods incorporating nonlinear parameter
     identification 190

9.3 Combined reconstruction and motion correction 192
    9.3.1 The advantages of the list mode format 193
    9.3.2 Motion correction during an iterative reconstruction
          algorithm 194
          9.3.2.1 Approaches based on a rigid or affine
                  motion model 194
          9.3.2.2 Approaches based on a non-rigid
                  motion model 196
9.4 Combination of parameter identification and
    motion estimation 198
References 200

III   Recent Developments   207

10 Introduction Hybrid Tomographic Imaging   209
Hartwig Newiger
10.1 Introduction 209
10.2 Combining PET and SPECT 210
10.3 The combination with MR 211
10.4 Combining ultrasound with PET and SPECT 214
References 215

11 MR-based Attenuation Correction for PET/MR   217
Matthias Hofmann, Bernd Pichler, and Thomas Beyer
11.1 Introduction 218
11.2 MR-AC for brain applications 220
    11.2.1 Segmentation approaches 220
    11.2.2 Atlas approaches 221
11.3 Methods for torso imaging 224
11.4 Discussion 229
    11.4.1 The presence of bone 230
    11.4.2 MR imaging with ultrashort echo time (UTE) 231
    11.4.3 Required PET accuracy 232
    11.4.4 Validation of MR-AC methods 232
    11.4.5 Truncated field-of-view 232
    11.4.6 MR coils and positioning aids 233
    11.4.7 User intervention 233
    11.4.8 Potential benefits of MR-AC 234
    11.4.9 Additional potential benefits of simultaneous
          PET/MR acquisition 234
11.5 Conclusion 234
References 235

12 Optical Imaging   **241**

Angelique Ale and Vasilis Ntziachristos

12.1 Introduction   241
12.2 Fluorescence molecular tomography (FMT)   244
    12.2.1 Light propagation model   244
        12.2.1.1 Photon interaction with biological tissue . .   244
        12.2.1.2 The diffusion approximation   246
        12.2.1.3 Model for a fluorescence heterogeneity   248
    12.2.2 Reconstruction of the fluorochrome distribution   249
12.3 FMT and hybrid FMT systems   251
    12.3.1 Instrumentation   251
        12.3.1.1 Illumination   251
        12.3.1.2 Detection   252
        12.3.1.3 360° projections   252
    12.3.2 Multimodal optical imaging   253
        12.3.2.1 Optical tomography and MRI   253
        12.3.2.2 FMT-XCT   254
References   257

Index   **263**

About the Series

The *Series in Medical Physics and Biomedical Engineering* describes the applications of physical sciences, engineering and mathematics in medicine and clinical research.

The series seeks (but is not restricted to) publications in the following topics:

- Artificial Organs

- Assistive Technology

- Bioinformatics

- Bioinstrumentation

- Biomaterials

- Biomechanics

- Biomedical Engineering

- Clinical Engineering

- Imaging

- Implants

- Medical Computing and Mathematics

- Medical/Surgical Devices

- Patient Monitoring

- Physiological Measurement

- Prosthetics

- Radiation Protection, Health Physics and Dosimetry

- Regulatory Issues

- Rehabilitation Engineering

- Sports Medicine

- Systems Physiology

- Telemedicine

- Tissue Engineering

- Treatment

The *Series in Medical Physics and Biomedical Engineering* is an international series that meets the need for up-to-date texts in this rapidly developing field. Books in the series range in level from introductory graduate textbooks and practical handbooks to more advanced expositions of current research.

The *Series in Medical Physics and Biomedical Engineering* is the official book series of the International Organization for Medical Physics.

The International Organization for Medical Physics

The International Organization for Medical Physics (IOMP), founded in 1963, is a scientific, educational, and professional organization of 76 national adhering organizations, more than 16,500 individual members, several Corporate Members and four international Regional Organizations.

IOMP is administered by a Council, which includes delegates from each of the Adhering National Organizations. Regular meetings of Council are held electronically as well as every three years at the World Congress on Medical Physics and Biomedical Engineering. The President and other Officers form the Executive Committee and there are also committees covering the main areas of activity, including Education and Training, Scientific, Professional Relations and Publications.

Objectives

- To contribute to the advancement of medical physics in all its aspects.

- To organize international cooperation in medical physics, especially in developing countries.

- To encourage and advise on the formation of national organizations of medical physics in those countries which lack such organizations.

Activities

Official journals of the IOMP are *Physics in Medicine and Biology, Medical Physics* and *Physiological Measurement*. The IOMP publishes a bulletin *Medical Physics World* twice a year which is distributed to all members.

A World Congress on Medical Physics and Biomedical Engineering is held every three years in co-operation with IFMBE through the International Union for Physics and Engineering Sciences in Medicine (IUPESM). A regionally based International Conference on Medical Physics is held between World Congresses. IOMP also sponsors international conferences, workshops and courses. IOMP representatives contribute to various international committees and working groups.

The IOMP has several programs to assist medical physicists in developing countries. The joint IOMP Library program supports 69 active libraries in 42 developing countries and the Used Equipment Program coordinates equipment donations. The Travel Assistance Program provides a limited number of grants to enable physicists to attend the World Congresses.

The IOMP website is being developed to include a scientific database of international standards in medical physics and a virtual education and resource centre.

Information on the activities of the IOMP can be found on its web site at www.iomp.org.

Foreword

In the past 40 years the field of tomography has evolved from simple X-ray projection imaging, autoradiography, fluorescent imaging and magnetic resonance imaging to production of the three dimensional images of the accumulation of chemical tracers injected into animals, plants and human subjects. The first major breakthrough technology since Röntgen's discovery of X-ray projection imaging was X-Ray computer-assisted tomography known in the 1970s as CAT scanning or just CT. The method allowed one to compute the internal distribution of tissue characteristics and injected contrast material from the amount of X-Rays (photons) transmitted through an object from multiple angles. Though analog methods were available to accomplish the representation of what is inside an object from projections by merely back projecting the beams onto photographic material, the digital computer-enabled X-Ray computer assisted tomography to revolutionize medical radiography. X-rays aimed in multiple directions through an object (patient) allowed measurements that could be used to calculate the most likely distribution of tissue parameters that would result in the data observed by detectors positioned at multiple angles around the object.

The mathematical problem is as follows. Given the number of X-Ray photons entering the object in a particular direction and given the number of photons that were able to exit the object, estimate the amount of material that attenuated the photons and do this estimation for many paths through the object. The problem is generally known as the inverse problem wherein an unknown is calculated from known observations that represent or are a transformation of the unknown. Mathematically this turns out to be a simple manipulation because the ratio between the number of photons from the X-Ray generator that enter the body at a particular angle to the number of photons exiting the body along a line from the X-ray generator to that detector can be converted to the sum of the individual pieces of tissue between the X-ray input and the detected output for a particular line through the object. The conversion was merely the logarithm of this ratio.

By accumulating these ratios from many projections of X-rays, one can answer the question: What is the most likely distribution of tissue attenuation that can give the observed results? The X-Ray CAT scan is the 3D distribution of attenuation which is generally equivalent to the distribution of electron density.

So if this problem of tomography was solved 40 years ago with the main

TABLE 0.1: Characteristics of X-Ray and Emission Tomography

X-Ray Tomography	Emission Tomography
Known	Known
Source strength	
Source position	
Detector response	Detector response
Detector position	Detector position
Unknown	Unknown
	Source strength
	Source position
Attenuation distribution	Attenuation distribution
	Diffusion/Scatter

progress being made in perfection of computer methods to achieve and display the information as well as methods to improve the collection of information, why this book? Indeed, even progress in removal of artifacts from motion such as respiration and heart motion and changes in photon energy spectra within the object have been dealt with, so why not use the X-Ray computed tomography methods for imaging injected radionuclides, radiopharmaceuticals and fluorescent molecules? The most concise answer is that emission tomography has four unknowns for a six-parameter problem but transmission tomography has only one unknown for a five-parameter problem.

In X-ray computed tomography the mathematical problem is to compute the amount of attenuation given the known input X-ray photons and the known number of output photons detected from the object at every angle or projection. But, in emission tomography the problem is to compute the distribution of photon sources inside the body without knowing the intensity of the sources, nor their position, nor the amount that is attenuated and the contamination of the detected photons from scattering elsewhere in the object. What is known is the amount of photons that get out of the object along a particular trajectory (see Table 0.1). Thus we must estimate the source strengths, their position, and the attenuation of the emitted photons as well as how many of the photons detected were scattered from multiple sources. To a mathematician this is an intractable problem (cf. *Phys. Med Biol* 19:387-389, 1974); nevertheless it is solved by methods discussed in this book along with methods to compensate for motion, partial volume, registration and other factors that influence detector performance.

It is refreshing to have a text on emission molecular imaging relevant to animals and human beings with an emphasis on those factors that detract from resolution and quantification. This book implicitly distinguishes between molecular imaging of emitters and molecular imaging provided by magnetic resonance techniques such as magnetic resonance spectroscopy, magnetic resonance imaging of hyperpolarized and other contrast agents, and other magnetic resonance methods wherein the response to the injected pattern of the radiofrequency field is measured. The mathematics of image reconstruction of an intrinsic emitter that is not stimulated by a known external probing signal are more complicated for light and gamma ray photon emitter reconstruction problems. The exception to this statement is functional molecular tomography (FMT) wherein the behavior of the stimulating photons and that of the excited photons from injected flurophors need to be incorporated in the reconstruction strategy (Chapter 12). Yet, the benefits of emission tomography for molecular imaging of radionuclide emissions relative to other modalities lie in the exquisite sensitivity of radionuclide detection and the broad scales in time, space and object size served by SPECT and PET techniques as well as their role in hypbid imaging systems that employ a combination of emission with CT, MRI and FMT.

In sum, this book shows how researchers have overcome limitations in emission tomography noted 40 years ago and have brought the methods to the goal of high spatial resolution and quantification. Most importantly, these advances have enabled clinically useful applications not available to other diagnostic methods.

Thomas F. Budinger, MD, PhD
Professor
University of California, Berkeley

List of Contributors

Angelique Ale
Institute for Biological and Medical
    Imaging
Technical University of Munich and
    Helmholtz Center Munich
Neuherberg, Germany

Martin Benning
Institute for Computational and
    Applied Mathematics
University of Münster
Münster, Germany

Thomas Beyer
cmi-experts GmbH
Zürich, Switzerland

Ralph A. Bundschuh
Rechts der Isar Hospital
Technical University of Munich
Munich, Germany

Florian Büther
European Institute for Molecular
    Imaging
University of Münster
Münster, Germany

Mohammad Dawood
European Institute for Molecular
    Imaging
University of Münster
Münster, Germany

Michael Fieseler
Department of Mathematics and
    Computer Science
University of Münster
Münster, Germany

Fabian Gigengack
Department of Mathematics and
    Computer Science
University of Münster
Münster, Germany

Sven Hermann
European Institute for Molecular
    Imaging
University of Münster
Münster, Germany

Matthias Hofmann
Laboratory for Preclinical Imaging
    and Imaging Technology
University of Tübingen
Tübingen, Germany

Marc Huismann
Department of Nuclear Medicine &
    PET Research
VU University Medical Center
Amsterdam, the Netherlands

Andreas H. Jacobs
European Institute for Molecular
 Imaging
University of Münster
Münster, Germany

Xiaoyi Jiang
Department of Mathematics and
 Computer Science
University of Münster
Münster, Germany

Thomas Kösters
European Institute for Molecular
 Imaging
University of Münster
Münster, Germany

Frederic Lamare
Medical Research Council Clinical
 Sciences Centre
Imperial College London,
 Hammersmith Campus
London, United Kingdom

Marilyn Law
European Institute for Molecular
 Imaging
University of Münster
Münster, Germany

Hartwig Newiger
Molecular Imaging
Siemens Healthcare
Erlangen, Germany

Vasilis Ntziachristos
Institute for Biological and Medical
 Imaging
Technical University of Munich and
 Helmholtz Center Munich
Neuherberg, Germany

Bernd Pichler
Laboratory for Preclinical Imaging
 and Imaging Technology
University of Tübingen
Tübingen, Germany

Klaus P. Schäfers
European Institute for Molecular
 Imaging
University of Münster
Münster, Germany

Michael Schäfers
European Institute for Molecular
 Imaging
University of Münster
Münster, Germany

Sonja Schäfers
European Institute for Molecular
 Imaging
University of Münster
Münster, Germany

Daniel Tenbrinck
Department of Mathematics and
 Computer Science
University of Münster
Münster, Germany

Thomas Viel
European Institute for Molecular
 Imaging
University of Münster
Münster, Germany

Frank Wübbeling
Department of Mathematics and
 Computer Science
University of Münster
Münster, Germany

Sibylle I. Ziegler
Rechts der Isar Hospital
Technical University of Munich
Munich, Germany

Chapter 1

Introduction

Klaus P. Schäfers

European Institute for Molecular Imaging, University of Münster, Münster, Germany

1.1	Introduction	1
1.2	Principle of emission tomography	2
1.3	Electromagnetic spectrum	4
1.4	Need for correction techniques	4
References	....	6

1.1 Introduction

"Seeing is believing." This idiom indicates that imaging plays an important role in human life and, in particular, in biomedical research and medical diagnostics. The detection of X-rays in 1895 by Wilhelm Conrad Röntgen changed the way of medical diagnostics radically. For the first time ever, physicians were offered a view inside the living body without the need for opening it by surgery. This very first imaging technology paved the way for modern patient diagnostics and treatment.

Since then, many imaging techniques have been exploited and developed for biomedical use based on different principles using either electromagnetic waves (gamma rays, X-rays, ultraviolet, visible, or infrared light) or magnetic resonance (magnetic resonance imaging, MRI).

This book is focused on different correction techniques used in emission tomography to generate and enhance images. In a general context, emission tomography is commonly defined as an imaging method that makes use of radioactive isotopes, e.g., applying techniques like single photon emission tomography (SPECT) or positron emission tomography (PET). We have extended the definition toward the general use of emitting probes based on electromagnetic radiation. Therefore, new optical imaging (OI) approaches that aim to visualize a three-dimensional volume using either luminescense or fluorescence probes are also encompassed in this book.

The process of image formation based on emission tomography necessitates a number of corrections. Images may become quantitative and allow inter- and

1

intra-subject comparison only after application of these corrections. For PET imaging, for instance, this includes corrections for radioactive decay, random and scattered events, corrections for partial volume effects, compensation for patient, heart and breathing motion, attenuation correction, dead time correction, etc.

Additionally, hybrid emission tomography (e.g., PET-CT, PET-MRI, SPECT-CT, PET-OI), which is increasingly being used due to the opportunities presented by hybrid imaging, necessitates another range of corrections, such as attenuation correction of PET data based on CT or MRI information.

This book deals with the required correction methods from the point of view of computer science, mathematics and physics. The objective is to present an overview of relevant problems in emission tomography, possible artifacts caused by them, and correction techniques to improve the resulting images.

1.2   Principle of emission tomography

Emission tomography is an imaging technique where a three-dimensional distribution of photons is emitted by sources such as radionuclides or fuorescent compounds that are injected into the biological or human object. These sources are part of the electromagnetic spectrum covering a large variety of waves with different wavelengths and properties.

In emission tomography the radiation sources are transferred into an object of interest (e.g., the human body or a mouse) by different procedures, in most cases by injection into the blood stream. After distribution, the injected substance (probe) emits electromagnetic waves which can then be detected from outside the object by dedicated detectors.

Unlike X-ray imaging where the images show the absorption of X-rays traveling through an object, emission imaging makes use of emitting probes that have the capability of penetrating tissue. Whereas in X-ray imaging contrast agents are often utilized to enhance the images, the emitting probes themselves build the contrast in emission imaging.

The most interesting but often most challenging task is the development of specific probes where the radiation-emitting substance is chemically bound (labeled) to a molecule of interest. After injection of this labeled compound into a biological system, one can follow (*trace*) noninvasively the labeled molecule of interest. This phenomenon is called the *tracer principle* which was first discovered by George Charles de Hevesy. Due to this behavior, emission tomography is often referred to as a *molecular imaging* technique. Consequently, emission images show molecular or functional information rather than morphological or structural details. This has led to the development of hybrid imaging techniques, such as PET-CT or SPECT-CT, which combine high-level molecular

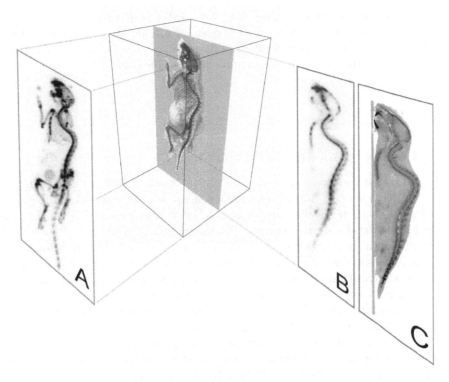

FIGURE 1.1: (See color insert.) Two different biomedical imaging principles demonstrated on a ^{18}F-fluorid PET bone scan of a mouse: (A) projection image showing the maximum intensity projection; (B) single PET slice out of the 3D tomographic volume; (C) same slice but fused with CT data.

with co-registered anatomical information, thus combining the best of the two imaging worlds.

Two different imaging principles have been developed in emission tomography: projection imaging and tomography.

Similar to the principle of a camera, a projection is showing a two-dimension picture of the tracer activity projected onto a surface outside the object. If our eyes were sensitive to gamma rays we could directly see the emitted "light" shining through the human body (or animal) as a projection image (Figure 1.1 A). Volumes of high tracer accumulation would appear bright, areas of low tracer uptake would barely be visible.

Tomography is an imaging technique where the biological object is shown as a stack of two-dimensional slices. Slices through the object of interest are visualized as if the observer would be looking directly onto these slices (Figure 1.1 B and C). In contrast to projection imaging, tomography is a three-dimensional imaging method showing details of the local tracer distribution.

Tomographic images are usually *reconstructed* during a mathematical process from many projection images acquired at certain angles around the biological object. As image reconstruction is essential for advanced emission tomography, a chapter of the principle of image reconstruction is included in this book.

1.3   Electromagnetic spectrum

Emission tomography is an imaging technique based on electromagnetic radiation, a phenomenon of oscillating self-propagating waves comprising electric and magnetic field components. Electromagnetic radiation can be classified into several subtypes according to the frequency or wavelength of its wave. Examples are radio waves, infrared radiation, visible light, ultraviolet radiation, X-rays and gamma rays. The human eye is sensitive to only a small band of the electromagnetic spectrum called the visible spectrum. The basic "unit" (quantum) of all electromagnetic waves is the *photon*.

Photons traveling through biological tissue may be absorbed and scattered by water, suspended particles, and dissolved matter. The attenuation coefficient is a measure of the conversion of energy to heat and chemical energy; thus, the higher the attenuation coefficient is, the higher the probability of photons to be absorbed or scattered by the medium.

Analyzing the attenuation coefficients at different wavelengths (Figure 1.2), it becomes obvious that gamma radiation has good properties for emission imaging since it is only mildly attenuated by water or biological tissue. The emitted gamma photons can be detected and localized outside the biological sample without being completely absorbed. Photons out of the visual light spectrum have also ideal properties in water, but the absorption becomes significant if hemoglobin is on the pathway of these lower energy photons. Since hemoglobin is always part of organs or biological tissue, optical imaging is feasible only at the level of a few centimeters rather than at higher ranges as compared to imaging with gamma rays.

1.4   Need for correction techniques

The generation of tomographic images involves a number of processing steps until image quality and accuracy is reached that is sufficient for visual inspection or further image-based analysis. In general, the workflow for emission tomographic imaging can be described by five processing units: data

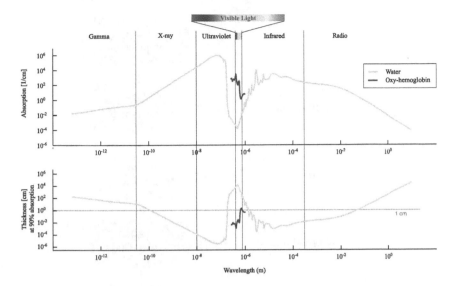

FIGURE 1.2: **(See color insert.)** The electromagnetic spectrum showing the absorption coefficient in water at different wavelengths. Both, gamma radiation and visible light have low attenuation in water; oxy-hemoglobin adds a considerable attenuation (compiled from [1, 2, 3]).

acquisition, data processing, image generation, image processing, and image analysis (Figure 1.3).

Since acquired data are always affected by physical limitations of the acquisition hardware, corrections have to be applied at a very early stage in order to improve data and correct for these effects. In PET and SPECT imaging, this involves correction for dead time, detector efficiencies, etc. In optical imaging light scattering in biological tissue is the main limitation for accurate reconstruction of planar or even tomographic images. Highly sophisticated reconstruction schemes and correction have to be applied to generate semi-quantitative images. Many of these corrections need to be applied at the level of data processing.

After data acquisition, the pre-corrected data are usually reconstructed by imaging reconstruction techniques aiming for generating either planar images or a three-dimensional tomographic volume. Since this is the most important processing step that defines and determines the quality of the final results, most of the corrections are directly integrated into the image reconstruction procedure. For PET (and partly for SPECT), corrections for scattered photons, decay, and attenuation are usually peformed during image reconstruction.

Finally, the reconstructed images may be post-processed to further enhance images and improve quantitative accuracy. Noise reduction by image filtering,

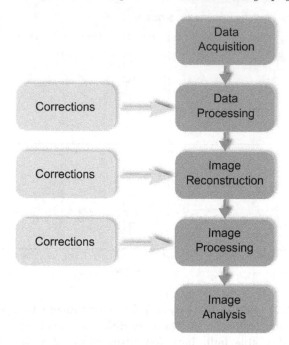

FIGURE 1.3: General workflow of emission tomographic imaging. Corrections have to be applied in order to get artifact-free and quantitative imaging results.

partial volume correction and motion correction are some examples of those post-processing procedures. Image quality may be significantly improved after applying these corrections allowing quantitative image analysis.

It is not exaggerating to claim that emission tomographic imaging would not be feasible without the use of corrections at the different stages of data and image processing. There is a need for dedicated correction techniqes in all imaging modalities. The better the corrections work, the better the image quality and accuracy will be.

This book gives a comprehensive overview of correction techniques at the different levels of the data processing workflow. Since there is a large number of correction methods available dedicated to individual scanner designs and hardware, this book is not meant to be complete in the description of all available methods, but will demonstrate the basic principles as well as some novel methods. A strong focus of this book is on motion correction techniques in PET imaging as this is a sophisticated and new approach towards high-resolution, high-quality PET imaging which will most certainly have a great impact on imaging analysis and image-based diagnostics.

References

[1] Molar extinction coefficients of oxy and deoxyhemoglobin—compiled by Scott Prahl, http://omlc.ogi.edu/spectra/hemoglob.

[2] X-ray mass attenuation coefficients—compiled by the National Institute of Standards and Technology (NIST), http://physics.nist.gov/physrefdata/xraymasscoef/tab4.html.

[3] D. J. Segelstein. The complex refractive index of water. PhD thesis, University of Missouri–Kansas City, 1981.

Part I

Background

Chapter 2

Biomedical Applications of Emission Tomography

Michael Schäfers, Sven Hermann, Sonja Schäfers, Thomas Viel, Marilyn Law, and Andreas H. Jacobs

European Institute for Molecular Imaging, University of Münster, Münster, Germany

2.1	The role of imaging in biomedical research and applications		11
2.2	Functional and molecular imaging by emission tomography enables high sensitivity and spatial resolution		13
2.3	Biomedical applications of emission tomography depend on tracers		14
2.4	Applications		16
	2.4.1	Preclinical applications	16
	2.4.2	Clinical applications	17
	2.4.3	Examples of biomedical applications of emission tomography	18
		2.4.3.1 Bioluminescence imaging of tumor growth ...	18
		2.4.3.2 Dynamic PET in pharmakodynamic studies .	19
		2.4.3.3 From mice to men—Non-invasive translational imaging of inflammatory activity in graft-versus-host disease	20
		2.4.3.4 PET to quantify catecholamine recycling and receptor density in patients with arrhythmias	22
		2.4.3.5 Multiparametric imaging of brain tumors ...	23
References		....	26

2.1 The role of imaging in biomedical research and applications

Over the past decades imaging has gained an increasingly important role in biomedical research and clinical decision making [30, 31]. This is mostly due to the specific strength and beauty of imaging, which is (in the majority of imaging techniques) the non-destructive multi-parametric characterization of a single cell, a tissue or complex organisms ranging from model organisms, such as drosophila, to patients. In this context imaging methods cover a broad spectrum from high resolution microscopic techniques over dedicated small animal scanners to whole-body human systems in clinical diagnostics.

The field of medical imaging had already been founded when, in 1895, the German physicist Wilhelm Conrad Röntgen discovered "invisible rays" which he initially termed *X-rays*. X-rays in combination with X-ray sensitive films were immediately applicable for planar imaging in medical diagnostics when used in a transmission approach. One of the first examples was clinical imaging of the thorax with the patient positioned in front of a film cassette and the X-ray tube in front of the patient. The rapid spread of X-ray-based imaging in medicine was especially driven by technological advances such as digital detectors, tomography (computer tomography) and contrast agents, which were important milestones in the entire field of medical diagnostics. X-ray imaging depicts the density of structures by their specific X-ray absorption; dense structures such as bones can be excellently imaged, whereas soft tissue contrast is limited. X-ray imaging is therefore primarily used for high resolution morphological imaging; contrast enhancement is used to delineate structures such as vessels.

With the discovery of radioactive radiation by Henry Becquerel in 1896 a new biomedical imaging principle was founded, which nowadays has become a major tool in biomedical imaging. Based on the initial description of radioactivity by Becquerel, George Charles de Hevesy first described in 1923 the use of radioactive labelling of molecules to trace their fate upon application into a biological system such as plants, cells, animal models, or even patients. With the parallel development of gamma detectors as the basis of scintigraphy replacing the Geiger-Müller-Counters, imaging of the distribution of radioactivity in organisms became a reality. These inventions essentially founded the field of diagnostic functional and molecular imaging in medicine. One of the earliest examples of a medical application of scintigraphic imaging was the first thyroid imaging study of Paul Blanquet in Bordeaux as early as 1951. In a patient with thyroid nodules Blanquet applied $[^{131}I]$, which, as the "original" non-radioactive iodine $[^{127}I]$, is taken up by the thyroid cells and reflects their molecular metabolic activity. In this case $[^{131}I]$ "traces" the fate of the physiological iodine in the thyroid gland since biochemically $[^{131}I]$ is identical to the "cold" $[^{127}I]$, $[^{131}I]$ being an example of an *authentic* tracer. Today, $[^{131}I]$ has been replaced by injection of $[^{123}I]$ or $[^{99m}Tc]O_4$; however, the principle of functional and molecular thyroid diagnostics is unchanged and was not replaced by any other technique.

2.2 Functional and molecular imaging by emission tomography enables high sensitivity and spatial resolution

Functional and molecular imaging aims at the visualization of molecular processes in complex biological systems ranging from synthetic molecular structures, from cells to organs to individuals. Innovative functional imaging technologies, namely the imaging of molecular processes on a single moleculular level, uniquely enable the exploration of chemical, physiological and pathophysiological processes such as receptor expression and function, catalytic (enzymatic) activities and cell-cell interactions in biosystems.

The specific strength of imaging lies in the ability to resolve dynamic processes in time and space in systems as small as ion transporters to large complex organisms such as patients suffering from a specific disease. The knowledge gained by imaging primarily supports the understanding of molecular interactions. This is important for basic research of biological functions but also, and most importantly, for future developments in medicine.

The diagnostic imaging of functional and molecular processes *in vivo* needs a high sensitivity of the imaging technology plus a reasonable spatial and temporal resolution tailored to the diagnostic question. Although the repertoire of clinical imaging techniques covers a whole range from morphological to functional to molecular imaging modalities, only the scintigraphic technologies, single photon emission tomography (SPECT) and positron emission tomography (PET), and optical imaging methods, such as bioluminescence and near-infrared fluorescence imaging, are based on emission tomography and have an exclusively high sensitivity.

As detailed in the previous chapter, scintigraphic techniques are preferable for deep tissue imaging since gamma rays can travel long distances through tissues without significant interferences. Therefore, SPECT and PET are mainly used for functional and molecular imaging in clinical algorithms. "Miniaturized" dedicated SPECT and PET systems have paved the way for application of scintigraphic approaches in preclinical research [20]. Nowadays, optical imaging using genetically encoded bioluminescence or injectable fluorescent dyes is used extensively in preclinical research. However, fluorescence imaging is just starting to be used for clinical imaging of structures on or close to the body surface or in a catheter-based setup [15]. Both emission-based techniques provide a very high molecular sensitivity in the nano- to picomolar range *in vivo*, enabling the quantification of receptors, enzymes, transmitters in organisms while having adequate spatial and temporal resolution. This exquisitely high sensitivity is crucial, since low "tracer" amounts of substances, which do not have a pharmacological effect and therefore do not influence the biosystems one is looking at, can be detected.

2.3 Biomedical applications of emission tomography depend on tracers

As already mentioned above, biomedical applications of emission tomography use the injection of labeled molecules to trace a functional aspect or molecular target and address specific scientific questions. In general, all of these applications are grouped into "molecular imaging," which is defined quite differently in the context of basic sciences vs. clinical diagnostics or natural sciences vs. medicine. In the literature, multiple definitions can be found; for example, "Molecular imaging can be defined as the *in vivo* characterization and measurement of biological processes at the cellular and molecular level...", "...implies the convergence of multiple image-capture techniques, basic cell/molecular biology, chemistry, medicine, pharmacology, medical physics, biomathematics, and bioinformatics into a new imaging paradigm." [12, 29]. These definitions point to the important aspect of interdisciplinary team work between biology, chemistry, mathematics, physics, computer sciences and medicine to successfully develop and apply innovative molecular imaging technologies. One of these aspects, the generation and optimization of images from emission raw data is within the focus of this book. Signals that can be reconstructed from emission data into image data stem from and depend on tracers, and their chemical development and pharmaceutical validation is an important step in molecular imaging. These tracers typically consist of two components:

(a) a **drug (pharmacophor)**, which either follows a functional principle and therefore traces it or targets specific molecular signatures (molecular targets). This molecule or drug compound of the tracer decides on its biodistribution. For example, a tracer for measuring tissue perfusion is typically based on a compound that is taken up by cells at a high first pass rate and trapped in the cell. The resulting emitted imaging signal therefore is quantitatively related to the degree of tissue perfusion. A classical example for molecular imaging is the quantification of the density of a receptor on the cell surface. This approach is typically realized by using receptor-affine molecules/drugs, such as a receptor antagonist, which specifically bind to a receptor or receptor family.

(b) a **flag (label)**, which is tightly bound to the drug component and is emitting signals. This flag can consist of radioactive isotopes such as positron emitters or gamma emitters allowing deep tissue penetration with a fair resolution or a fluorescent probe having a limited application for deep tissues but allowing micro- to nanoscopic resolution.

Upon injection of the drug-flag complex the fate of the drug can be followed by its emission signals with crystal detectors, CCD cameras, etc., allowing the non-invasive visualization and quantification of molecular targets or functional parameters addressed by the drug (Figure 2.1).

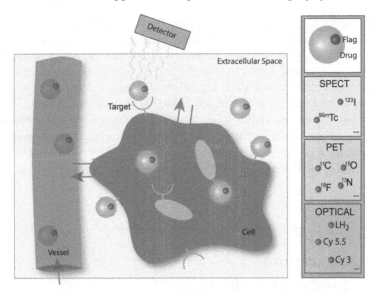

FIGURE 2.1: (See color insert.) The tracer principle. Molecular targets are typically addressed and visualized by injection of tracers, which are consisting of a drug (light green) to which a flag (dark green; radioactivity, fluorescent dyes, quenched optical dyes, etc.) is attached. The tracer arrives in organs via blood vessels, diffuses into the extracellular space and can bind to externalized targets (orange; receptors, etc.) on the cell surface or cross the cell membrane to bind to intracellular targets. By its flag the tracer emits light or gamma rays to be detected from outside the organism by SPECT, PET or optical imaging. Examples of isotopes and dyes used for SPECT, PET and optical imaging are listed.

Many of the tracer approaches make use of amplification strategies to optimize the target-to-non-target ratio. A prominent example is the tracer fluorine-18-2-deoxy-2D-glucose (FDG) which is the most common and clinically established [^{18}F] labeled tracer. It suffices for many applications, with the majority of studies performed for tumor imaging and imaging of inflammatory and neuronal pathologies as well as imaging of cardiovascular diseases. Since FDG is a glucose analogue its uptake by cells is correlated to the rate of glycolysis. Upon intravenous injection FDG follows the initial biochemical route of glucose being taken up intracellularly by glucose transporters. As glucose, FDG is phosphorylated by hexokinase to FDG-6-P in the cell. However, in contrast to glucose, FDG-6-P is not further metabolized since it does not undergo metabolism in the citrate cycle. As a result FDG is trapped intracellularly. Since a single hexokinase enzyme can react "serially" with many FDG molecules, the imaging signal is amplified.

The identification of targets specific for a scientific or clinical question is the first and crucial step in the development of a tracer. These targets can be located "behind" barriers and therefore might not be easily accessible upon injection of a tracer into the blood. In this context, such barriers are the cellular membrane, which does not allow for a general entrance of drugs, and the blood-brain barrier being highly selective to protect the brain. Beside this barrier aspect the accessibility of a molecular target for a given tracer *in vivo* is dependent on the structure of the compound (antagonist, agonist, enzyme inhibitor, etc.), which should provide high affinity and specificity while having low metabolism. Reviews are suggested for further reading [2, 7, 27, 17].

2.4   Applications

2.4.1   Preclinical applications

With the development of tailored and dedicated imaging devices for studying animals such as small animal PET and SPECT or fluorescence reflectance and tomography, the field of applying emission tomography in preclinical studies has broadened substantially over the past years. Recent developments have achieved high resolution small animal imaging and now do allow human-like imaging even in mice with respect to sensitivity, and temporal and spatial resolution. This is of great interest for translational studies using PET and SPECT, which have been established for years in clinical algorithms, whereas optical imaging is limited in this respect. Since in small animal scanners the imaging technology is pushed to its theoretical limit and the size of the object to be imaged is small, artifacts conflicting with accurate and quantitative imaging can occur from various angles (partial volume, attenuation, movement, etc.) and demand the application of dedicated correction methods.

Small animal imaging can be and frequently is used in different aspects of basic and translational biomedical research such as:

- *Phenotyping*. Imaging to characterize mouse models mimicking human diseases. Phenotyping of new and existing animal models is substantially supported by functional and molecular small animal imaging, especially since small animal imaging characteristically can be performed non-invasively and serially in individual animal. Furthermore, whole-body imaging is of particular value to discover pathologies developing in organs "outside" the focus of the study.

- *Monitoring*. Imaging to monitor morphological, functional and molecular changes in the spontaneous time course or induced by interventional studies (pharmaceutical treatment, surgical intervention, gene therapy, stem cell therapy). Again, serial non-invasive imaging studies enable assessment of intra-individual changes in a few animals.

- *Development.* Since the majority of clinical imaging devices are now available in a "miniaturized" fashion for small animal imaging with similar imaging characteristics, it seems straightforward to use small animal imaging in validation studies to characterize new imaging approaches in animal models first and then translate into clinical imaging by using the analogous clinical imaging modality. This is especially useful when testing new imaging probes for molecular imaging (contrast agents, radiopharmaceutical, optical dyes, etc.).

2.4.2   Clinical applications

In current medicine, much of the medical practice is based on standards of care that are determined by averaging responses across large cohorts. Based on clinical trials every patient receives the same treatment, biological variability within diseases and between individuals is only sparsely considered. On the contrary, biomedical imaging can assess the individual patient's specific characteristics on the level of individual molecular signatures locally and quantitatively in patients' whole body. This is expected to be superior to systemic analyses such as blood tests when looking for the clinically important local burden of diseases. Biomedical imaging uniquely allows doctors to detect hidden or early disease, to select and tailor appropriate therapies, and to monitor therapy efficiency in individuals. In this respect—given its large potential for studying physiology and pathophysiology non-invasively in humans and patients—emission tomography has seen widespread application over the last 20 years in clinical research and diagnostics. With the advent and expansion of the field "molecular imaging" emission tomography per se was pushed to higher levels with optical imaging most frequently used in preclinical and basic research studies. The reasons are the low costs of the technique, easy labeling strategies of fluorescent dyes, the long half-life and the link to microscopy. However, optical imaging in humans is still and will most likely always be restricted to applications where the imaging object is close to the surface (skin, endoscopy, etc.) due to light scattering and absorption when light travels through deep tissues. On the contrary, SPECT and PET have been established for many years for clinical diagnostics and research and founded the success of clinical nuclear medicine. SPECT and PET are not limited to surface applications and can quantitatively sense radioactive signals in the whole body. A huge library of clinical and experimental radiopharmaceuticals is available. While SPECT tracers are mainly produced in-house by applying the gamma emitter ^{99m}Tc (6 h half-life) eluted from ^{99}Mo generators to commercial kits, PET relies on the cyclotron-based production of short-living (typical range from 2 min to 2 h half-life) isotopes coupled to precursors in automated synthesis procedures at the site of the application or within short distance to it.

The molecular imaging with SPECT and PET has successfully entered clinical algorithms in various fields of oncological, cardiovascular, neurological and other types of diseases. Applications range from pathophysiological studies to clinical diagnostics, therapy control and prevention.

While clinical SPECT imaging relies on a broad variety of radiopharmaceuticals (perfusion agents, metabolism tracers, receptor ligands, etc.) the success of PET is based on the use of [18F]FDG which can be used quite universally in various conditions since it reflects glucose metabolism, a key metabolic pathways in organisms. However, the glucose signal is relatively unspecific for a single disease entity; for example, macrophages accumulate [18F]FDG in inflammatory lesions while growing tumors do the same. The development of more specific tracers therefore remains a huge challenge for radiochemistry.

2.4.3   Examples of biomedical applications of emission tomography

Since the field of biomedical applications is broad and quickly expanding, this chapter aims at showing relevant areas of applying emission tomography from recent work of our group rather than giving a complete overview. For further reading current reviews are suggested. A special emphasis is put on challenges for quantification of emission tomography [12, 23, 1, 24, 4, 25, 28, 8, 18, 3].

2.4.3.1   Bioluminescence imaging of tumor growth

Bioluminescence imaging (BLI) is a non-invasive, sensitive, rapid and cost-effective means to investigate cellular behavior in a living organism. Although the typical image acquisition is planar resulting in 2D images, the technique can be used to quantify signals from species such as zebra fish or mice. Signals within cells are typically generated by the use of reporter genes. A prominent reporter gene is the firefly luciferase of *Photinus pyralis*. This enzyme catalyzes the oxidation of D-luciferin in an ATP-dependent manner resulting in the emission of light (emission maximum ∼560 nm). To follow the fate of cells in organisms, specific cell populations are transfected with the firefly luciferase gene. One example is the follow-up of cancer growth and response to therapy by the use of transduced cells expressing luciferase. Upon injection into animals the transduced tumor cells proliferate and divide but still carry the luciferase gene. Bioluminescence imaging makes use of the intravenous injection of luciferin, which is taken up by cells and only in the case of luciferase expression is oxidized in a chemical reaction that results in the emission of light. The emitted bioluminescence allows non-invasive following of the *in vivo* growth or regression of the genetically modified tumor cells reflecting viable tumor cell mass. Although the image acquisition is planar, a direct correla-

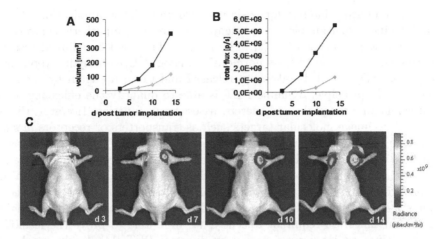

FIGURE 2.2: **(See color insert.)** Longitudinal BLI analysis of the growth of subcutaneous brain tumors. CD1 nu/nu mice were injected twice with $1*10^6$ human U87ΔEGFR glioma cells that stably express firefly luciferase. At different days post tumor implantation the tumor volume was measured with the help of a caliper (A); tumor volume $= 0.52 \times$ length $\times$ width^2). Furthermore, tumor activity was assessed by BLI measurements 10 min after intraperitoneal injection of 2 mg D-luciferin using the IVIS spectrum system (Caliper Life Sciences). Serial BL images displayed in pseudocolors superimposed to a white light image show tumor growth of two subcutaneous tumors at the left and right back region (C). BLI signals that arise from the tumors were quantified (B) (gray: left tumor, black: right tumor) to assess viable tumor volumes.

tion of luciferase expression with tumor volume as measured by, for example, caliper measurement [16] or MRT [19], has been observed. Figure 2.2 shows the use of the luciferase reporter gene for studying the spontaneous growth of subcutaneous tumors in individual mice over time. Precise algorithms for translating 2D bioluminescence signals into measures of cell numbers and providing 3D images remain a huge challenge in optical imaging due to scatter and attenuation of light in tissue.

2.4.3.2 Dynamic PET in pharmakodynamic studies

A unique feature of PET in small animals is the dynamic assessment of radioactivity distribution in the whole animal to assess pharmakodynamics. An example of using this approach to assess quantitatively the nerve function in murine hearts is given here. For the function of the heart, the sympathetic nervous system is as important as the perfusion or myocardial contractility. In heart failure and arrhythmias, activity of the sympathetic nervous system is impaired; the quantitative study of nerve function at the heart is therefore of great interest for basic research and clinical application. Several radioligands are available to investigate the cardiac sympathetic nervous system in patients using PET [9]. The most widely used for clinical research is the catecholamine ("stress hormone") analog $[^{11}C]meta$-hydroxyephedrine ($[^{11}C]m$HED), which is taken up, released and taken up again by the nerve endings. In principle, dynamic PET is able to quantify the efficiency of the catecholamine re-uptake and thereby assess nerve function. However, the use of PET to image molecular targets such as transporters or receptors in small animals imposes challenges not apparent in studies using metabolic tracers such as $[^{18}F]$FDG which can be given at the high concentrations needed to achieve good images. Scanner design aims to optimize both resolution and sensitivity but in dedicated animal scanners resolution is often pursued at the expense of sensitivity so that high doses of radioactivity and/or long acquisition times are required. Recently, we were able to prove the feasibility of assessing cardiac nerve function by dynamic PET [11]. In this study, the radiolabelled catecholamine $[^{11}C]m$HED was injected intravenously into mice and its pharmakokinetics followed by dynamic PET acquisitions. From the mathematical analysis of the time-activity-curves generated from regions-of-interest placed over the left ventricular myocardium, measures of the efficiency of the catecholamine uptake were calculated. The great potential of dynamic PET is shown by injecting an uptake competitor of $[^{11}C]m$HED, metaraminol after the injection of $[^{11}C]m$HED and while the animal was scanned dynamically in PET. Competetion of $[^{11}C]m$HED binding by metaraminol can be quantitatively assessed by the analysis of $[^{11}C]m$HED washout from the myocardium (Figure 2.3). Since the object of interest, the murine heart, is small and beating fast, correction methods for partial volume, spillover, motion, etc. are of crucial importance to guarantee an optimal precision of quantification.

2.4.3.3   From mice to men—Non-invasive translational imaging of inflammatory activity in graft-versus-host disease

As stated above, $[^{18}F]$FDG is the most commonly used tracer for PET imaging in patients. Therefore, applying it in small animal PET to study the diagnostic potential of a new application and translating the preclinical findings into clinical research is a obvious and natural approach. An example of this translational strategy is given here.

Gastrointestinal graft-versus-host disease (GvHD) is a common and potentially life-threatening complication after hematopoietic stem cell transplantation (HSCT). Non-invasive tests for assessment of GvHD activity are desirable but lacking. Since GvHD is an inflammatory condition with inflammatory cells infiltrating various organs such as the bowel and $[^{18}F]$FDG is known to be taken up by inflammatory cells, we aimed at testing the ability to visualize intestinal GvHD-associated inflammation in an allogeneic murine transplant model by $[^{18}F]$FDG-PET *in vivo* [26]. A predominant localization of intestinal GvHD to the colon was verified by histology and fluorescence reflectance

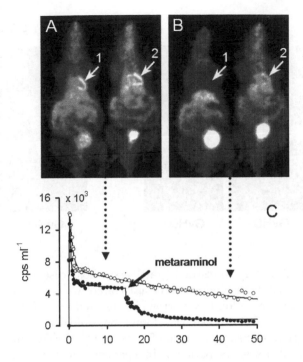

FIGURE 2.3: Dynamic PET imaging of the sympathetic nervous system in two mice following simultaneous i.v. injection of [^{11}C]mHED. (A) shows PET images with good uptake of the tracer in the heart (arrows) in both mice 10 min p.i. Mouse 1 was additionally injected with metaraminol 15 min p.i. (B) shows PET images 30 min after application of metaraminol (40 molkg^{-1} i.v.) in mouse 1, where there is virtually no [^{11}C]mHED uptake left in the heart of mouse 1 while the heart of mouse 2 shows only a slight washout. (C) depicts time-activity-curves generated from regions-of-interest placed over the heart and calculated from the PET listmode stream (• mouse 1, ○ mouse 2). Note the spontaneous slight washout of [^{11}C]mHED from the myocardium in the untreated mouse 2 but the dramatic loss of the tracer upon injection of the competitor metaraminol in mouse 1. From the tracer dynamics, measures for efficiency of neuronal re-uptake can be calculated. Modified from [11]

imaging of enhanced green fluorescent protein (EGFP) expressing donor cells. Infiltration of colon tissue by EGFP-positive donor lymphocytes matched increased FDG uptake in serial PET examinations in GvHD-positive mice. Most interestingly, in a translational approach, these preclinical data were confirmed in a cohort of patients with suspected intestinal GvHD, where the intestinal FDG uptake was highly predictive for clinically relevant GvHD (Figure 2.4). This study is a prime example for the translational potential of PET but also

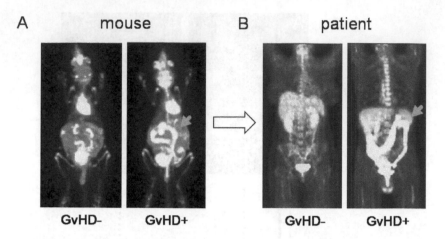

FIGURE 2.4: (A): Intestinal graft-versus-host disease (GvHD+) in mice after stem cell transplantation is associated with increased local glucose metabolism in the colon (gray arrow) as assessed by [18F]FDG-PET, whereas intestinal [18F]FDG uptake was found physiological in the control group (GvHD-). (B): The same pattern of intestinal [18F]FDG uptake of the mouse model was also seen in patients. Independent of symptoms GvHD-positive patients exhibited a strong intestinal [18F]FDG uptake as compared to gvHD-negative patients with physiological [18F]FDG uptake of the bowel. Modified from [26].

emphasizes the need for common quantification strategies between preclinical and clinical imaging.

2.4.3.4 PET to quantify catecholamine recycling and receptor density in patients with arrhythmias

Although less established in clinical practice than perfusion imaging, examination of the cardiac autonomous nervous system is of potential value. This may apply especially to arrhythmogenic diseases not associated with functional and anatomic changes detectable by conventional imaging and may also be useful in ischemic heart disease where aberrations in autonomous nervous function may be a relevant parameter.

Both pre- and postsynaptic function of the sympathetic and parasympathetic nervous system are accessible by radiopharmaceutical techniques. At present, the sympathetic arm has received most attention. The investigation of the presynaptic function (catecholamine release and recycling) of sympathetic innervation by the SPECT radiotracer [123I]-meta-iodobenzylguanidine or the PET radiotracer [11C]-meta-hydroxyephedrine ([11C]HED) has led to new insight into the role of cardiac nerves in various cardiovascular disorders (see above).

Apart from the presynaptic innervation, the postsynaptic density of β adrenoceptors to which catecholamines bind and whereby they mediate intracellular responses is an important parameter. β adrenoceptors have been quantified by using radiolabeled β adrenoceptors antagonists and PET. [^{11}C]CGP 12177 is a non-selective β adrenoceptor antagonist that has seen the most widespread application in clinical research [22, 14].

PET, with its capability to quantify transmitter synthesis and transport as well as adrenoceptor density non-invasively *in vivo*, needs sophisticated tracer-kinetic modelling. In the field of PET radiopharmaceuticals for assessment of cardiac sympathetic nervous function, two tracer-specific models exist. Presynaptic norepinephrine re-uptake function is assessed by calculation of the distribution volume (V_d) of ^{11}C-hydroxyephedrine using a single compartment model and least square non-linear regression analysis to provide influx and efflux rate constants [22]. Delforge et al. [5] described a model for measurement of myocardial adrenoceptor density (B_{max}) with a double injection protocol of ^{11}C-CGP 12177, which implies two injections of different amounts of radioactivity and a cold substance, using a graphical approach. The main advantage of this approach is that the results are obtained without having to measure the input function and without estimating the metabolites. With this model and technique, the quantification of the active adrenoceptors located on the surface of the plasma membrane can be achieved *in vivo* since ^{11}C-CGP 12177 has hydrophilic characteristics and therefore does not cross the cell membranes significantly. Modelling of tracer kinetics to quantify receptor densities and nerve function is critically dependent on precise correction methods for emission tomography data taking into account attenuation, extravascular volume fractions and such. Figure 2.5 demonstrates the quantification of individual presynaptic and postsynaptic sympathetic innervation through kinetic modelling of two dynamic PET studies in patients with different forms of arrhythmias.

2.4.3.5 Multiparametric imaging of brain tumors

Gliomas are the most common primary brain tumors with an incidence of 5-10/100.000. Gliomas still carry a very limited prognosis and together with all intra-cranial neoplasms they are the second most common cause of death from an intracranial disease after stroke. In recent years imaging based on MRI and PET has considerably improved the management of patients with gliomas. The most common PET markers implemented are [^{18}F]FDG for glucose consumption and cellular density, [^{11}C]-*methyl*-methionine (MET) for amino acid transporter activity and neovascularisation, and [^{18}F]-fluoro-L-thymidine (FLT) for cellular thymidine kinase activity and tumor cell proliferation. All imaging markers aim toward the characterization of the biological activity of the tumor (Figure 2.6; [6]). In the clinical application this is espe-

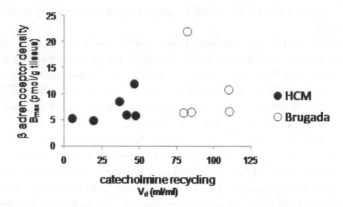

FIGURE 2.5: Absolute quantification of cardiac presynaptic (volume of distribution of [^{11}C]mHED, X axis) and postsynaptic sympathetic innervation (B_{max} of [^{11}C]CGP 12177, Y axis) in patients with arrhythmias using dynamic PET and tracer-kinetic modelling. Although the images look homogeneous and similar between patient groups, there are significant quantitative differences in cardiac innervation between patients with hypertrophic cardiomyopathy (HCM) and Brugada syndrome (Brugada) that are detectable only by tracer-kinetic modeling.

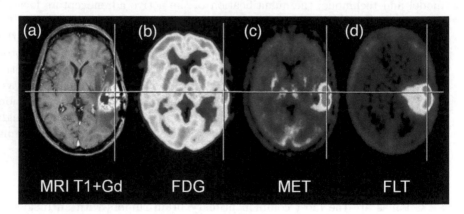

FIGURE 2.6: **(See color insert.)** MRI and PET are being used together to assess (a) the breakdown of the blood-brain barrier (T1+Gd); (b) the metabolic activity of the tumor as assessed by [^{18}F]FDG-PET, which also serves as a surrogate marker for cellular density; (c) the uptake of radiolabelled amino acids such as [^{11}C]MET, which serves as direct marker for the expression of amino acid transporters and as a surrogate marker for neovascularization; (d) the uptake of radiolabelled thymidine ([^{18}F]FLT), which serves as direct marker for cellular thymidine kinase activity and as surrogate marker for cell proliferation. The various imaging markers give complementary information on the activity and extent of the tumor.

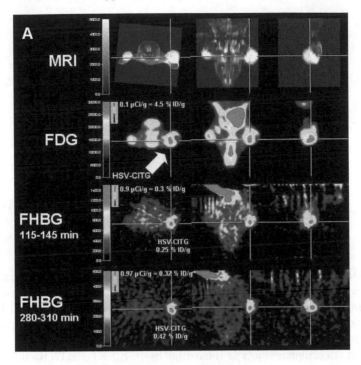

FIGURE 2.7: (See color insert.) Imaging-guided gene therapy paradigm of experimental gliomas. Protocol for identification of viable target tissue and assessment of vector-mediated gene expression *in vivo* in a mouse model with three subcutaneous growing gliomas. Row 1: localization of tumors by MRI. Row 2: the viable target tissue as depicted by [18F]FDG-PET. Note the signs of necrosis in the lateral portion of the left-sided tumor (arrow). Rows 3–4: following vector-application into the medial viable portion of the tumor (arrow) the "tissue-dose" of vector-mediated gene expression is quantified by [18F]FHBG-PET. Row 3 shows an image acquired early after tracer injection, which is used for coregistration. Row 4 displays a late image with specific tracer accumulation in the tumor that is used for quantification.

cially important to guide biopsy, resection and radiation as well as to assess the effect of therapy and rate of tumor progression.

The same imaging parameters have been used extensively in the past by our group in experimental glioma models to further develop an imaging-guided gene therapy paradigm, where vectors are transduced into the viable tumor parts as identified by PET (Figures 2.7 and 2.8, [10]). The transduced tissue dose of vector-mediated gene expression can by visualized by PET and can be correlated to the induced therapeutic effect. We also observed that FLT can serve as an early (within 4 days after onset of therapy) read-out parameter for

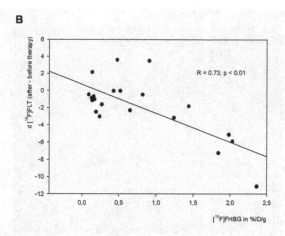

FIGURE 2.8: Imaging-guided gene therapy paradigm of experimental gliomas. Response to gene therapy correlates to therapeutic gene expression. The intensity of therapeutic gene expression (cdIREStk39gfp), which is equivalent to transduction efficiency and the "tissue-dose" of vector-mediated therapeutic gene expression, is measured by [^{18}F]FHBG-PET (in %ID/g), and the induced therapeutic effect is measured by [^{18}F]FLT-PET (R=0.73, p<0.01). The therapeutic effect ([^{18}F]FLT) was calculated as the difference between [^{18}F]FLT accumulation before and after therapy.

anti-proliferative gene therapeutic effects [21]. The same imaging paradigms that are being used to image transduced genes by PET or optical imaging can be applied to study stem cell transplantation [28] or transcriptional regulation in gliomas [13].

References

[1] E. O. Aboagye. The future of imaging: developing the tools for monitoring response to therapy in oncology: the 2009 Sir James MacKenzie Davidson Memorial Lecture. *The British Journal of Radiology*, 83(994):814–822, 2010.

[2] G. Antoni and B. Langstrom. Radiopharmaceuticals: molecular imaging using positron emission tomography. *Handbook of Experimental Pharmacology*, (185 Pt 1):177–201, 2008.

[3] B. R. Arenkiel and M. D. Ehlers. Molecular genetics and imaging technologies for circuit-based neuroanatomy. *Nature*, 461(7266):900–7, 2009.

[4] K. Chang and F. Jaffer. Advances in fluorescence imaging of the cardiovascular system. *Journal of Nuclear Cardiology*, 15(3):417–428, 2008.

[5] J. Delforge, A. Syrota, J. P. Lancon, K. Nakajima, C. Loc'h, M. Janier, J. M. Vallois, J. Cayla, and C. Crouzel. Cardiac beta-adrenergic receptor density measured in vivo using PET, CGP 12177, and a new graphical method. *Journal of Nuclear Medicine*, 32(4):739–748, 1991.

[6] F. G. Dhermain, P. Hau, H. Lanfermann, Andreas H. Jacobs, and Martin J. van den Bent. Advanced MRI and PET imaging for assessment of treatment response in patients with gliomas. *The Lancet Neurology*, 9(9):906–20, 2010.

[7] W. C. Eckelman, R. C. Reba, and G. J. Kelloff. Targeted imaging: an important biomarker for understanding disease progression in the era of personalized medicine. *Drug Discovery Today*, 13(17-18):748–759, 2008.

[8] D. A. Hammoud, J. M. Hoffman, and M. G. Pomper. Molecular neuroimaging: from conventional to emerging techniques. *Radiology*, 245(1):21–42, 2007.

[9] M. M. Henneman, F. M. Bengel, E. E. van der Wall, J. Knuuti, and J. J. Bax. Cardiac neuronal imaging: application in the evaluation of cardiac disease. *Journal of Nuclear Cardiology*, 15(3):442–455, 2008.

[10] A. H. Jacobs, M. A. Rueger, A. Winkeler, H. Li, S. Vollmar, Y. Waerzeggers, B. Rueckriem, C. Kummer, C. Dittmar, M. Klein, M. T. Heneka, U. Herrlinger, C. Fraefel, R. Graf, K. Wienhard, and W.-D. Heiss. Imaging-guided gene therapy of experimental gliomas. *Cancer Research*, 67(4):1706–15, 2007.

[11] M. P. Law, K. Schäfers, K. Kopka, S. Wagner, O. Schober, and M. Schäfers. Molecular imaging of cardiac sympathetic innervation by 11C-mHED and PET: From man to mouse? *Journal of Nuclear Medicine*, 51(8):1269–1276, 2010.

[12] T. F. Massoud and S. S. Gambhir. Molecular imaging in living subjects: seeing fundamental biological processes in a new light. *Genes & development*, 17(5):545–580, 2003.

[13] P. Monfared, A. Winkeler, M. Klein, H. Li, A. Klose, M. Hoesel, Y. Waerzeggers, S. Korsching, and A. H. Jacobs. Noninvasive assessment of E2F-1-mediated transcriptional regulation in vivo. *Cancer Research*, 68(14):5932–40, 2008.

[14] M. Naya, T. Tsukamoto, K. Morita, C. Katoh, K. Nishijima, H. Komatsu, S. Yamada, Y. Kuge, N. Tamaki, and H. Tsutsui. Myocardial beta-adrenergic receptor density assessed by 11C-CGP12177 PET predicts improvement of cardiac function after carvedilol treatment in patients with idiopathic dilated cardiomyopathy. *Journal of Nuclear Medicine*, 50(2):220–225, 2009.

[15] V. Ntziachristos. Going deeper than microscopy: the optical imaging frontier in biology. *Nature Methods*, 7(8):603–614, 2010.

[16] Z. Paroo, R. A. Bollinger, D. A. Braasch, E. Richer, D. R. Corey, P. P. Antich, and R. P. Mason. Validating bioluminescence imaging as a high-throughput, quantitative modality for assessing tumor burden. *Molecular Imaging*, 3(2):117–124, 2004.

[17] S. L. Pimlott and A. Sutherland. Molecular tracers for the PET and SPECT imaging of disease. *Chemical Society Reviews*, 40(1):149–62, 2010.

[18] M. E. Raichle. Two views of brain function. *Trends Cogn Sci*, 14(4):180–90, 2010.

[19] A. Rehemtulla, L. D. Stegman, S. J. Cardozo, S. Gupta, D. E. Hall, C. H. Contag, and B. D. Ross. Rapid and quantitative assessment of cancer treatment response using in vivo bioluminescence imaging. *Neoplasia*, 2(6):491–495, 2000.

[20] D. J. Rowland and S. R. Cherry. Small-animal preclinical nuclear medicine instrumentation and methodology. *Seminars in Nuclear Medicine*, 38(3):209–222, 2008.

[21] M. A. Rueger, M. Ameli, H. Li, A. Winkeler, B. Rueckriem, S. Vollmar, N. Galldiks, V. Hesselmann, C. Fraefel, K. Wienhard, W.-D. Heiss, and H. Jacobs. [(18)F]FLT PET for non-invasive monitoring of early response to gene therapy in experimental gliomas. *Molecular Imaging and Biology*, 2010.

[22] M. Schäfers, D. Dutka, C. G. Rhodes, A. A. Lammertsma, F. Hermansen, O. Schober, and P. G. Camici. Myocardial presynaptic and postsynaptic autonomic dysfunction in hypertrophic cardiomyopathy. *Circulation Research*, 82(1):57–62, 1998.

[23] U. Schnockel, S. Hermann, L. Stegger, M. Law, M. Kuhlmann, O. Schober, K. Schäfers, and M. Schäfers. Small-animal pet: a promising, non-invasive tool in pre-clinical research. *European Journal of Pharmaceutics and Biopharmaceutics*, 74(1):50–54, 2010.

[24] M. E. Seaman, G. Contino, N. Bardeesy, and K. A. Kelly. Molecular imaging agents: impact on diagnosis and therapeutics in oncology. *Expert Reviews in Molecular Medicine*, 12:e20, 2010.

[25] S. Y. Shaw. Molecular imaging in cardiovascular disease: targets and opportunities. *Nature Reviews Cardiology*, 6(9):569–579, 2009.

[26] M. Stelljes, S. Hermann, J. Albring, G. Kohler, M. Loffler, C. Franzius, C. Poremba, V. Schlosser, S. Volkmann, C. Opitz, C. Bremer, T. Kucharzik, G. Silling, O. Schober, W. E. Berdel, M. Schäfers, and J. Kienast. Clinical molecular imaging in intestinal graft-versus-host disease: mapping of disease activity, prediction, and monitoring of treatment efficiency by positron emission tomography. *Blood*, 111(5):2909–2918, 2008.

[27] W. Wadsak and M. Mitterhauser. Basics and principles of radiopharmaceuticals for PET/CT. *European Journal of Radiology*, 73(3):461–469, 2010.

[28] Y. Waerzeggers, M. Klein, H. Miletic, U. Himmelreich, H. Li, P. Monfared, U. Herrlinger, M. Hoehn, H. H. Coenen, M. Weller, A. Winkeler, and A. H. Jacobs. Multimodal imaging of neural progenitor cell fate in rodents. *Molecular Imaging*, 7(2):77–91, 2008.

[29] R. Weissleder and U. Mahmood. Molecular imaging. *Radiology*, 219(2):316–333, 2001.

[30] R. Weissleder and M. J. Pittet. Imaging in the era of molecular oncology. *Nature*, 452(7187):580–589, 2008.

[31] J. K. Willmann, N. van Bruggen, L. M. Dinkelborg, and S. S. Gambhir. Molecular imaging in drug development. *Nature Reviews Drug Discovery*, 7(7):591–607, 2008.

Chapter 3

PET Image Reconstruction

Frank Wübbeling

Department of Mathematics and Computer Science, University of Münster, Münster, Germany

3.1	Introduction	31
3.2	Analytical algorithms	32
	3.2.1 Mathematical basis	32
	3.2.2 Filtered backprojection	35
	3.2.3 Implementation: Resolution and complexity	37
	3.2.4 Implementation and rebinning	38
	3.2.4.1 2D Rebinning	39
	3.2.4.2 3D filtered backprojection	40
	3.2.5 Limitations	40
3.3	Discrete algorithms	40
	3.3.1 ART—Algebraic reconstruction technique	41
	3.3.2 EM	42
	3.3.3 Computing the system matrix	44
	3.3.4 List mode	45
3.4	Summary	47
References	....	47

3.1 Introduction

In this chapter, we give a short review of image reconstruction algorithms for positron emission tomography. The basic problem is: How do we convert the measurements of counts on a set of lines of response into a plot of the activity function f of a radioactive agent?

Basically, we distinguish two types of algorithms: analytical algorithms based on a mathematical analysis of the Radon transform, usually employing the filtered backprojection, and numerical algorithms based on a discretization of the problem, usually employing the expectation-maximization algorithm or one of its variants. While the former one is very fast and mathematically rigorous in two dimensions, it does not have an obvious extension to 3D, nor can it easily handle deviations to the Radon transform such as scatter, geometrical deficiencies or missing data. On the other hand, discretization algorithms are typically hard to handle numerically, since they involve the inversion of big matrices, but the flexibility makes up for that deficiency, making them the standard algorithm for high–quality images to date.

We do not aim for an exhaustive coverage of the subject, but rather lay the groundwork for many other articles in this book. The existing literature is vast, including the following references.

For a detailed classical mathematical introduction to medical image reconstruction, we refer the reader to [10]. For an approach oriented toward current algorithms, see [11]. For a classical engineering approach, see [8]. For a more up-to-date engineering approach, see [2]. For a historical overview of inversion formulas, see [9].

3.2   Analytical algorithms

In the most basic model of PET, the number of detected events on a line of response (LOR) is proportional to the amount of radioactivity on that line. Mathematically, that is given by the line integral over the activity distribution function f over the LOR. The fixed proportionality coefficent is derived from the total number of counts and the measurement time. Since the object under investigation is finite, we can assume that f has compact support Ω.

Let us assume for the moment that measurements for all lines passing through Ω are available, and that everything is 2D. Our problem can then be restated as follows:

Given measurements

$$m_L = \int_L f(x)\,\mathrm{d}x$$

for all lines L though Ω, compute f from theses measurements.

This is the classical formulation of the Radon inversion problem, solved by Radon in 1910.

In this chapter, we leave out all mathematical delicacies, in particular with respect to appropriate function spaces, and all proofs. More exact formulations can be found in the provided literature.

3.2.1   Mathematical basis

Denote by $\mathcal{S}^{n-1}$ the unit ball in $\mathbb{R}^n$. Then for fixed $\theta \in \mathcal{S}^{n-1}$, $s \in \mathbb{R}$, the set $L(\theta, x) = \{x \in \mathbb{R}^n : x\theta = s\}$ is a hyperplane in $\mathbb{R}^n$ with normal vector θ and distance $|s|$ to the origin. In particular, if $n = 2$, $L(\theta, s)$ is a line perpendicular to θ, with signed offset s from the origin.

For a sufficiently smooth, fast–decaying function f in $\mathbb{R}^n$ we define the Radon transform as the integral over $L(\theta, s)$

$$Rf(\theta, s) = \int_{L(\theta,s)} f(x)\,\mathrm{d}x$$

and its companion, the X–ray transform, as the line integral over the line in direction θ passing through x

$$Pf(\theta, x) = \int_{\mathbb{R}} f(x + t\theta)\,\mathrm{d}t, \ x \in \mathbb{R}^n, \ x \in \theta^\perp.$$

Note that for $n = 2$, we have $Rf(\theta, s) = Pf(\theta^\perp, s\theta)$, so both are equivalent. For $n = 2$, both Radon and X–ray transforms are the mathematical realization of the physical PET measurement process, for $n = 3$, X–ray transform gives the correct model.

We define the backprojection operator R^* as

$$(R^*g)(x) = \int_{\mathcal{S}^{n-1}} y(\theta, x\theta)\,\mathrm{d}\theta$$

or $(R^*g)(x)$ the integral of g over all hyperplanes passing through x. Backprojection is the simplest approach for doing inversion on measured data: in order to recover the activity in a point x, simply average all measurements over lines through x. We will analyze this approach later; a simple example is in Figure 3.1.

Backprojection for P is defined accordingly for line integrals

$$(P^*h)(x) = \int_{\mathcal{S}^{n-1}} h(\theta, x')\,\mathrm{d}\theta$$

where x' is the projection of x onto $\theta^\perp$.

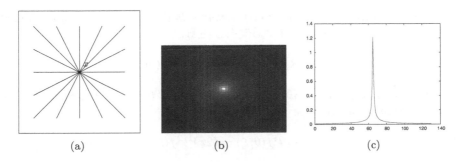

(a) (b) (c)

FIGURE 3.1: Unfiltered backprojection. (a) Integrate over all line integrals through x. (b) Simple unfiltered backprojection of a point source. (c) Point source backprojection cross section.

For $x, \xi \in \mathbf{R}^n$ we define the Fourier transform $\widehat{f}$ by

$$\widehat{f}(\xi) = (2\pi)^{-n/2} \int_{\mathbf{R}^n} f(x) e^{-ix\xi} \, \mathrm{d}x$$

and its inverse $f = \overset{\smile}{\widehat{f}}$ by

$$f(x) = (2\pi)^{-n/2} \int_{\mathbf{R}^n} \widehat{f}(\xi) e^{ix\xi} \, \mathrm{d}\xi.$$

The convolution of two functions f and g is defined by

$$(f * g)(x) = \int_{\mathbf{R}^n} f(y) g(x-y) \, \mathrm{d}y,$$

and we have the convolution theorem

$$\widehat{(f * g)}(\xi) = (2\pi)^{n/2} \widehat{f}(\xi) \widehat{g}(\xi).$$

Its main use is visible in its corollary of deconvolution: Assume that instead of a data function $g(x)$, we can observe only a function $h(x)$ as a convolution of g with a known function $\chi(x)$. Then, at least mathematically, we can recover $g(x)$ via

$$g(x) = \left(\widetilde{\widehat{h}() / \widehat{\chi}()} \right)(x).$$

Essentially, the theorem says that if rather than a true signal we measure a smoothed version of it, we can recover the original, provided the transfer function is known.

Note, however, that this will work in practice only when $|\widehat{\chi}| > \epsilon$ for a reasonable ϵ, which it usually is not, so this will usually not be implementable out of the box.

Now we have all the tools for the projection slice theorem.

Theorem 3.2.1 (Fourier Slice) *Assume that f is a fast decaying, smooth function on $\mathbf{R}^n$. Then, for $\theta \in \mathcal{S}^{n-1}$, $\sigma \in \mathbf{R}$,*

$$\widehat{Rf}(\theta, \sigma) = (2\pi)^{(n-1)/2} \widehat{f}(\sigma\theta)$$

where $\widehat{Rf}$ is a Fourier transform with respect to the second variable. For the X-ray transform, we have

$$\widehat{Pf}(\theta, \xi) = (2\pi)^{1/2} \widehat{f}(\xi)$$

where $\xi \perp \theta$, and $\widehat{Pf}$ is a $(n-1)$–dimensional Fourier transform in the second variable with respect to $\theta^{\perp}$.

Essentially, the theorem says that when we measure $Rf(\theta, s)$ for all θ and s, a 1D–Fourier transform with respect to the second variable will provide us with the Fourier transform of f, and a further nD–inverse Fourier transform will produce our function f. In particular, the Radon inversion problem is uniquely solvable.

While the theorem is easily proved and written, it is hard to implement. Suppose that Rf is measured on an equidistant grid in θ and s. Then, $\hat{f}$ is given on a polar rather than rectangular grid, which makes it impossible to use plain FFT for the inversion. While very efficient algorithms for the implementation of Fourier slice exist (see, e.g., [6]), usually the filtered backprojection approach is employed in PET.

3.2.2   Filtered backprojection

For simplicity, we assume that f is a 2D function. As a motivation for the filter algorithms, let us look at a simple interpretation of one of its variants first.

Assume that the distribution function f_z is the characteristic function of an arbitrarily small neighborhood Ω of a point z, meaning that all radioactivity is enclosed in a very small region around z. Thus, only line integrals passing through Ω will have a nonzero value.

The simplest idea for inversion of the Radon transform is that the value of $f(x)$ has an impact on $Rf(\theta, s)$ only if the corresponding line passes through x. Thus, we expect to get an approximation $f'(x)$ to $f(x)$ by simply averaging over all line integrals going through x. Mathematically, that amounts to evaluating $(R^*Rf)(x)$.

While f' is in fact an approximation to f, it is very blurred (see Figure 3.1). The reason for that is easily seen when we refer to f_z. f'_z will have its maximal value at z, but it will be not be zero outside of Ω since for every point x we find some lines going through x and Ω, so $Rf > 0$ for these lines, so the average value will not be zero. Rather, a simple geometrical argument shows that f'_z decays like $c/||x - z||$.

When we move z, f'_z is moved appropriately, so the mapping from f to f' is translation invariant, which implies that it is a convolution; this is already obvious from Figure 3.1. The convolution function is the response to a single peak at the origin, which we identified as $\chi(x) = c/||x||$. So, we have

$$(R^*Rf)(x) = (\chi * f)(x).$$

According to the corollary of the convolution theorem and using that in 2D, the Fourier transform of $1/r$ is $1/r$, we finally find that

$$\hat{f}(\xi) = \frac{1}{4\pi}||\xi|| \cdot \widehat{R^*Rf}(\xi).$$

Basically, this implies that R^*Rf is a smooth version of f, and all we have to do to regain f is to apply an appropriate edge–enhancing filter, the filter is defined in Fourier space by $||\xi||$. The algorithm goes by the name of ρ–filtered layergram, where ρ of course stands for the norm of ξ.

While this is perfectly implementable, a variant of this algorithm is usually employed. Instead of performing the filter step on the image delivered by the backprojection, it can be pulled through the operator R^* and performed directly on the data. Thus, the inversion formula now reads

$$f(x) = \frac{1}{4\pi}(R^*h)(x)$$

where

$$\widehat{h}(\theta, \sigma) = |\sigma|\widehat{Rf}(\theta, \sigma)$$

and the Fourier transform again refers to the second variable only.

Since this is the main theorem of the chapter, we take the liberty to give the very short proof:

$$
\begin{aligned}
\frac{1}{4\pi}(R^*h)(x) &= \frac{1}{4\pi}\int_{S^1} h(\theta, x\theta)\mathrm{d}\theta \\
&= \frac{1}{2\sqrt{2\pi}}\int_{S^1}\int_{\mathbf{R}} \widehat{h}(\theta, \sigma)e^{i\sigma x\theta}\mathrm{d}\sigma\mathrm{d}\theta \\
&= \frac{1}{2\sqrt{2\pi}}\int_{S^1}\int_{\mathbf{R}} |\sigma|\widehat{Rf}(\theta, \sigma)e^{ix\sigma\theta}\mathrm{d}\sigma\mathrm{d}\theta \\
&= \frac{1}{4\pi}\int_{S^1}\int_{\mathbf{R}} |\sigma|\widehat{f}(\sigma\theta)e^{ix\sigma\theta}\mathrm{d}\sigma\mathrm{d}\theta \\
&= \frac{1}{2\pi}\int_{\mathbf{R}^2} \widehat{f}(\xi)e^{ix\xi}\mathrm{d}\xi \\
&= f(x)
\end{aligned}
$$

where we have used the Fourier slice theorem and changed variables from polar to rectangular coordinates. The algorithm for filtered backprojection thus reads:

1. For a fixed measurement direction θ, compute the data's Fourier transform $\widehat{f}$.

2. Multiply $\widehat{f}(\xi)$ by $|\xi|/(4\pi)$.

3. Compute the backprojection of the result.

An equivalent formulation can be derived for the X-ray transform in 3D.

$$
\begin{aligned}
f(x) &= \frac{1}{(4\pi)^2}\int_{S^2} h(\theta, x')\mathrm{d}x = \frac{1}{(4\pi)^2}(P^*h)(x) \\
\widehat{h}(\theta, \xi) &= ||\xi||\widehat{(Rf)}(\theta, \xi)
\end{aligned}
$$

where again x' is the orthogonal projection of x on $\theta^\perp$. Note that the integration runs over all lines through x.

The algorithm for a point source is in Figure 3.2.

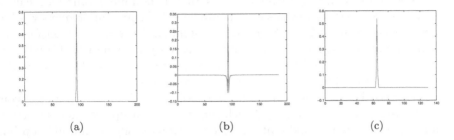

(a)   (b)   (c)

FIGURE 3.2: Filtered backprojection of a point source. (a) Measured data for a point source in one direction. (b) Same, with the Ram–Lak filter applied. (c) Point source backprojection cross section.

3.2.3 Implementation: Resolution and complexity

Note that the inversion of the Radon transform is an ill–posed problem, implying that small measurement errors can lead to arbitrarily large errors in the reconstruction. However, the degree of ill–posedness for $n = 2$ is $1/2$ on a Sobolev scale, with 1 being the degree of ill–posedness of the first derivative. In other words, applying the Radon transform twice in a row would be only as difficult, with respect to measurement errors, as taking the first derivative. So, we can expect reasonable reconstructions even from mildly polluted data.

However, we still have to take ill–posedness into account. In the backprojection algorithm, the source of ill–posedness is easily seen: high frequency-oscillations in the data are multiplied by the absolute value of the frequency, resulting in arbitrarily large errors. Consequently, the filter function is set to zero away from a limit frequency Ω. This can be done using a sharp cutoff (ramp) filter (cf. Ram-Lak filter) or in a smooth way, resulting in a selection of filters with more or less comparable output. We thus arrive at the final formulation for the filtered backprojection formula

$$f_\Omega(x) = (R^*h)(x)$$
$$\widehat{h}(\theta, \sigma) = \chi(\sigma)|\sigma|\widehat{Rf}(\theta, \sigma)$$
$$\chi(\sigma) = 0 \text{ for } |\sigma| > \Omega$$

where we choose $\chi(\sigma) = 1$, $|\sigma| \leq \Omega$ for the Ram–Lak filter.

The implementation is obvious; simply use the same algorithm as before, but set the value of the Fourier transform to zero beyond Ω.

However, this means that the output of our reconstruction formula is no longer exact, and can be only an approximation: since, obviously, h is bandlimited, f_Ω is, too. So, our reconstruction formula can be exact, that is $f_\Omega = f$, only for bandlimited functions f with band limit Ω.

Shannon's sampling theorem states that bandlimited functions can be uniquely reconstructed from samples, provided the sampling frequency is better than the Nyquist limit. Since for bandlimited f the Radon transform Rf is bandlimited too (using the projection theorem), we expect that a filter frequency of Ω corresponds to optimal sampling schemes of Rf that ensure that all Ω–bandlimited functions can be uniquely reconstructed.

Turning that argument around: given a scanner with fixed sampling geometry, there is a corresponding Ω such that filtered backprojection is exact for all functions f with bandlimit Ω, which gives an easily computable optimal choice for Ω, depending on scanner geometry only. Note that filtered backprojection thus has no more free parameters; everything can be chosen in an optimal way.

With respect to complexity, the backprojection step is dominating. Assuming that all parameters (resolution, number of data in each dimension, ...) are on the order of N, R^*h gives an image of size N^2 and needs N operations for the discretized integral, so the complexity of the overall algorithm is on the order of N^3. Note that using fast backprojection, this can be cut down to $O(N^2 \log N)$ operations (see, e.g., [13] for an introduction and multiple references). The same order can be achieved by using implementations of the Fourier slice theorem [6].

3.2.4 Implementation and rebinning

In the measurement process, single coincident photons on lines of response are recorded and counted. In order to satisfy the assumption that the number of photons on an LOR is proportional to the line integral over the activity function, we may need to aggregate lines of response into bins: in a measurement system of infinite accuracy, the probability of measuring two events on the same LOR is 0, so we would have a data vector of zeros and ones which does not satisfy our assumption. Note that this is the starting point of list mode algorithms.

Further, in the algorithm given above, we will be able to make use of only 2D-data, corresponding to a single ring of detectors, perpendicular to the axial direction of the detector in investigation. In order to make use of measurements in slant directions, we need to incorporate these into the reconstruction process.

There is no pure mathematical reason for doing this; using infinite measurement time, and infinite dose, the approximation of the data as an integral over activity function becomes exact, so there is no need for data from slant directions. However, of course, in an application we will try to make use of as many detected events as possible to enhance the signal to noise ratio.

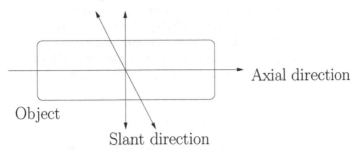

FIGURE 3.3: Single slice rebinning: Measurements of slant directions are used as additional measurements in the straight reconstruction plane.

3.2.4.1 2D Rebinning

Assume that our object is constant in the axial direction. Then, up to a constant factor, the integrals over lines that share the same projection on the transaxial plane will give the same measurement.

In this case, the single slice rebinning algorithm SSRB will be exact (see Figure 3.3):

1. Compute approximations to the 2D–Radon transform $Rf(\theta, s, z')$ of the activity function f on the plane $z = z'$ in the following way: For a reconstruction in a plane E perpendicular to the axial direction, we take into account the measurements g_L over all lines L that have their center point in E in the following way: Compute the projection of L onto the plane and use g_L as an additional measurement on E.

2. Reconstruct using 2D filtered backprojection. Obviously, this will not generally be exact for functions with variations in the z coordinate. In order to derive an exact formula, we first observe that if f is a 3D–function, Pf is a 4D–function. Thus, we expect the range of P to satisfy a one-dimensional consistency condition. One formulation is easily derived from the Fourier slice theorem for Pf: If θ, θ' are in $\mathcal{S}^{n-1}$, and ξ is in the intersection of $\theta^\perp$ and $\theta'^\perp$, we find that

$$\widehat{Pf}(\theta, \xi) = (2\pi)^{1/2}\widehat{f}(\xi) = \widehat{Pf}(\theta', \xi)$$

which basically means that it is possible to convert data measured in a given plane to measurement data for a different plane.

In fact, [3] proves that using an appropriate consistency condition, the measurement values for straight directions can be computed from the values for any slant direction. This is the starting point for Fourier rebinning type algorithms: Using approximations to the equation, measured values on slant lines are attributed to measurement values on transaxial planes. The main difficulty is that to implement the Fourier transforms involved, Pf must be available everywhere. Therefore, missing data are computed from the measurements for $\delta = 0$. For details of the complex implementation, see [4].

The dominating term with respect to complexity is a 3D Fourier transform with interpolation, which makes Fourier rebinning not too much slower than SSRB.

3.2.4.2   3D filtered backprojection

This is simply an implementation of the 3D X-ray backprojection formula. Note that to implement the corresponding algorithm, we need all lines passing through our object, but only a small number will actually be available. To get these, usually a low–quality preliminary reconstruction like SSRB is performed, and the missing data is taken from projections of the resulting image. The algorithm itself is on the order of N^5.

3.2.5   Limitations

The main advantages of the analytical methods are their speed and the fact that they can be fully analyzed mathematically, so exact bounds on resolution can be given and optimal choice of parameters is guaranteed. However, they are inflexible with respect to changes to the physical model and to the incorporation of non-Gaussian noise.

3.3   Discrete algorithms

Assume that the activity function f can be written (or at least approximated) as a linear combination of a finite number of ansatz–functions χ_k such that $f = \sum_{k=1}^{N} \alpha_k \chi_k$ with unknown coefficients α_k. Typically, χ_k are translates of each other, and could be Gaussians or voxels/pixels (characteristic functions of cubes or rectangles). For simplicity, we assume that the latter is the case.

Our continuous problem is thus reduced to the discrete problem of finding N numbers, satisfying the measurements. Assuming that data g_l are available for lines of response L_l, $l = 1 \ldots M$, our problem reads:

Find pixel values α_k, such that the line integral over L_l of the image is g_l:

$$
\begin{aligned}
g_l &= \int_{L_l} f(x)\mathrm{dx} \\
&= \int_{L_l} \sum_k \alpha_k \chi_k(x)\mathrm{dx} \\
&= \sum_k \int_{L_l} \chi_k(x)\mathrm{dx}\alpha_k \\
&=: \sum_k a_{lk}\alpha_k
\end{aligned}
$$

or

$$
g = A\alpha
$$

where $A = (a_{lk})$ is an M by N matrix, $\alpha = (\alpha_l)$ and $g = (g_k)$ are vectors. In the case of pixels, a_{lk} is just the length of the intersection of L_l with pixel k.

All we have to do is compute the system matrix A once and for all for a given system and solve the linear equation for each dataset g. Since A is extremely sparse, it is not economical to compute the (pseudo) inverse of A directly, and iterative methods are the appropriate tool for inversion.

Notice the big flexibility of these algorithms: valid for 2D and 3D, can deal with any scanner geometry, can easily incorporate additional measurements (cf. time of flight, TOF), and can incorporate information about f by choosing the χ_k appropriately. The major disadvantage is the usually slow runtime when compared to analytical algorithms.

3.3.1 ART—Algebraic reconstruction technique

ART is used for CT rather than PET, but we will gain some insight also for the EM algorithm.

In each iterative algorithm, we start with an initial guess α and update that value in each step, hopefully producing a sequence that converges toward a solution f. The main idea in the ART or Kaczmarz algorithm is to use only a small portion of the equations in each step. If the number of equations is small, the corresponding linear system can easily be inverted and used for the update.

The idea of ART or Kaczmarz can then be summarized as:

1. Start with an initial guess for α, typically, $\alpha = (0)$.

2. Choose one single measurement line L with measurement value g_L. Change the values of α such that the integral over L in the image becomes g_L.

3. Go back to 2.

Obviously, in step 2, we have an infinite number of choices for the change. We select the vector for α that makes the smallest (minimum-norm) change. Under mild conditions, this simple algorithm can be shown to converge.

Put in a more mathematical fashion, and using f_k for the image in the kth step of the algorithm:

1. Start off with an initial guess f_0, typically, $f_0 = 0$, and set $j = 0$.

2. Choose M' of the available equations. Extract the corresponding lines from A, g to A_j, g_j, such that $A_j f = g_j$.

3. Choose f_{j+1} as the orthogonal projection of f_j onto the subspace of all f with $A_j f = g_j$, that is:

$$f_{j+1} = f_j + \omega A_j^*(A_j A_j^*)^{-1}(Af_j - g)$$

   where A_j^* is the adjoint of A_j.

4. $j = j + 1$ and iterate from 2.

where we have introduced an iteration parameter ω. In order to circumvent the matrix inversion, $(AA^*)^{-1}$ can be replaced by any positive definite matrix C_j. We thus arrive at the final update step

$$f_{j+1} = f_j + \omega A_j^* C_j^{-1}(A_j f_j - g_j). \tag{3.1}$$

The iteration can be shown to converge towards the minimum norm solution of $Af - g$, provided ω is small enough, and all equations are used for the same number of times.

We remark that the convergence speed depends heavily on the order of selected equations. If in the first two steps of the algorithm, perpendicular matrix rows a_k are used, the algorithm gives the correct answer after only two iterations. However, if for any k, a_{k+1} is almost parallel to a_k, the improvement per step will be very small, and the algorithm will converge very slowly on the order of $1/k$. This remark is also true for N large.

So, the implicit rule should be: choose the equations in such a way that the corresponding matrix rows are as orthogonal as possible. While algorithms for achieving such an optimal order exist, it turns out that the main rule is to avoid the worst case of almost parallel lines. Choosing the equations at random usually shows not much difference in speed to optimal orderings.

3.3.2   EM

To fix ideas, we restrict our discretization model to 3D and voxels in this case.

Up to now, we completely neglected the fact that the data we have are measurements of a random rather than continuous process. In fact, the number of particles N_k emitted from a given voxel V_k and detected in the system, is a random variable. Obviously, its values are integers, and its distribution is given by the Poisson distribution

$$p(N_k = j) = e^{-f_k} \frac{f_k^j}{j!}.$$

The parameter f_k is the average number of particles emitted and thus proportional to the activity in voxel V_k which we want to compute. Note that in typical applications the number of decays measured on a single line is typically a small integer. This makes it crucial to take the noise model into account.

Denote by a_{lk} the probability that a particle is detected on LOR L_l, provided it has been emitted in voxel V_k (note that in an ideal scanner without scatter this definition matches the one given above, it is simply a re–interpretation). Then, the number of events g_k detected on line of response L_l is again a Poisson–distributed random variable, with mean value $(Af)_l$. The probability of measuring a given data vector $\mathbf{g}$, provided the activity distribution is given by f and all measurements are independent, is thus given by

$$P_{\mathbf{g}}(f) = \prod_l \frac{(Af)_l^{\mathbf{g}_l}}{\mathbf{g}_l!} e^{-(Af)_l}. \qquad (3.2)$$

Since we are looking at a random process, a delta source located at x could produce a measurement on an arbitrary line of response L. However, the probability is much higher if L passes through x. So the main idea is that for a given $\mathbf{g}$, find an f such that $P_{\mathbf{g}}(f)$ is maximized or find the distribution f that makes it most likely that $\mathbf{g}$ is measured. Consequently, f is denoted the maximum–likelihood solution.

In order to compute it, we define the EM algorithm [5], [16]

$$f_{j+1} = f_j \cdot \left(\frac{1}{A^t \mathbf{1}} A^t \frac{\mathbf{g}}{Af_j} \right)^\omega \qquad (3.3)$$

where $\omega > 0$ is an iteration parameter, $\mathbf{1}$ is a vector of 1s, and all multiplications and divisions of vectors are pointwise.

We immediately notice that for each (positive) f that satisfies $Af = \mathbf{g}$, f is a fixpoint of the iteration. Also, assuming that the equations are weighted such that $A^t \mathbf{1} = \mathbf{1}$, and replacing all vector products by summations, all subtractions by divisions, we get back the ART algorithm (3.1) with $A = A_j$. So, EM is nothing but the multiplicative version of ART.

It was shown in [16] that, under mild conditions, the EM algorithm converges toward the maximum likelihood solution. Its main drawback is that convergence is very slow. Since we already saw the relation of ART to EM, we can transfer Kaczmarz's main idea also to the EM algorithm: use (3.3) not for the full system of equations, but, as for the ART algorithm, select only a sub–system $A_l f = \mathbf{g}_l$. Thus we come up with the OSEM algorithm [7]

$$f_{j+1} = f_j \cdot \left(\frac{1}{A_j^t \mathbf{1}} A_j^t \frac{\mathbf{g}_j}{A_j f_j} \right)^\omega .$$

OSEM stands for ordered subset EM. As in the ART algorithm, again there is an optimal order of equations which ensures an optimal convergence rate. However, again as in ART, it turns out that it is usually sufficient to avoid the worst case by simply randomizing the equation selection process, so the true name of the algorithm should be OSEM in current implementations, and definitely in list mode.

Other than filtered backprojection, which implicitly contains a smoothing filter, EM does not contain a smoothing term. In a long iteration, typically high–frequency ("checkerboard") effects come up. The most effective way of dealing with them is to stop the iteration early, however, details may get lost this way. Many variants of the EM algorithm, making up for that deficiency, have been proposed, most of them incorporating penalty terms in 3.2. See [15] for an overview.

3.3.3 Computing the system matrix

In iterative methods, there is no need to compute the system matrix A explicitly. Since it is needed only when applied to a vector, typically, the matrix elements are computed on–the–fly. In the simple interpretation, when a_{lk} is the intersection of a line with a voxel, its value can be computed very quickly using the Siddon algorithm ([17], [1]). Note, however, that since detectors are not points but have a spatial dimension, the acceptance range of a line of response is actually approximately a cylinder.

Up to this point, we have neglected (at least) two properties of the underlying model. Due to the positron range, we do not measure the activity distribution f directly, but the distribution of photon emissions $f' = Df$, where D is a convolution. On the other hand, due to scatter between detectors, even $g' = Af'$ is not measured directly, but only $g = Eg'$, where again E is a convolution. So finally, our equation actually reads [14]

$$g = EADf$$

and Equation 3.3 has to be applied with A replaced by EAD. Since E and D are convolutions, they are conveniently implemented.

The interpretation of a_{lk} as the probability of a pixel from voxel k to be measured on line l has many interesting applications and allows incorporation of all linear effects. For a given PET system, the values might be computed once and for all, for example by a complex and computationally expensive Monte Carlo simulation. Without the need to understand and quantify the physical processes, a_{lk} can even be measured directly by placing a point source in voxel k and performing a full measurement. If done on a system-by-system basis, this is used to make up for deviations in a series of scanners and typically drastically enhances reconstruction away from the center.

Also, the incorporation of additional information like TOF poses no problem: TOF δ simply changes the probability of the origin of a recorded event on line L from homogeneous to Gaussian and can be easily incorporated into the system matrix, in particular for list mode (see below).

Note, however, one serious drawback apart from computation time. Many parameters in iterative algorithms (like resolution, stopping time, iteration parameters) can hardly be chosen in an optimal way and are usually estimated, as opposed to analytical methods with a full mathematical analysis that allows perfect choices.

3.3.4 List mode

For three reasons, rebinning is a nasty process. First, the position of particles and thus the line that defines a decay can be measured with high precision. In the rebinning process, this line is approximated by a line in S and the quality of measurement is not fully exploited. Second, to start the reconstruction process, we need to complete the rebinning process first, so there is a time delay—no just-in-time processing is possible. Third, the remarks above show that a completely random arrangement of equations is usually better than a structured one. By rebinning, we change the optimally random arrangement of the list of events that comes into a structured list (which can, of course, then be randomized again).

List mode solves these problems. List mode refers to the idea that equations are processed according to the **list of events** that comes directly from the detectors, without a rebinning process. The algorithm is most easily understood by theoretically choosing S to be the set of all lines rather than a discrete set of lines. Since decays are supposed to take place randomly with random direction, the probability that the same line is measured twice is zero. So g_L is zero for almost all L in S, except for those where an event was actually measured on L, in this case

$$g_L = 1.$$

Looking into the EM algorithm, we find that if a vector component is zero, the corresponding row in the matrix A^t does not contribute to the result, so in the computation of A^t we can safely delete all lines in A which belong to entries in g that vanish. We end up with a new system matrix A_{LM} and the algorithm

$$f^{k+1} = f^k \cdot \frac{1}{A^t \mathbf{1}} A^t_{LM} \frac{1}{A_{LM} f^k}. \tag{3.4}$$

It is very important to note here that the normalization factor is still A rather than A_{LM}. Of course, following this idea, the normalization factor has to be computed using the interpretation above rather than explicitly; applying $\mathbf{1}$ to the infinite–dimensional matrix would make no sense.

Also note that the algorithm is different from a competing approach which is tempting as a simple motivation: for each measured line L, write the equation

$$\int_L f(x)\mathrm{d}s = 1,$$

discretize and apply EM to the resulting system. This would result in the same algorithm as (3.4), with the exception that the normalization matrix would have to be chosen as A_{LM} as well. The simple example of a delta source at the origin shows that this makes no sense.

At first glance, this approach (choosing a very fine grid of lines) seems to break the interpretation of g as an approximation to the X-ray transform, since we have only 1s and 0s in g. However, this is not true: assume that in an experiment classical EM is applied and all lines measured are members of S, meaning that no precision is lost in rebinning. Note that this is the case with classical crystal sensor elements where data is naturally prediscretized since only a finite number of directions can be measured. Suppose further that line $L \in S$ has been awarded g_L events. In EM, we need to compute

$$A^t \frac{g}{Af^k}.$$

We construct a matrix A_{EM} which consists of the rows of A the row belonging to line L in A is written g_L times in A_{EM}. Then the update above just reads

$$A_{EM}^t \frac{1}{A_{EM}f^k}.$$

Since A_{EM} differs from A_{LM} only in the order of rows, in this case, EM list mode is exactly the same as classical EM. Note that this is valid only for EM, not for OSEM-like methods. In OSEM, the arrangement of the matrix is crucial, so it is not the same thing.

Since EM list mode is just a special case of EM, all convergence acceleration methods like OSEM, optimal choice of convergence parameters and so on can and must be applied.

However, there is a serious drawback, which of course is computation time. The size of system matrix A is much bigger than in the classical rebinned system; we get one equation per measured event. This is in part accounted for by an optimal randomization of events: Not only are the single rebinned equations written in a random order, but from the rebinning point of view even the equations are split into single equations with $\mathbf{1}$ on the right hand side and randomized. So a major reduction of number of iterations can be expected and is usually seen.

List mode statistics have been investigated thoroughly in papers by Parra and Barrett [12].

3.4 Summary

Analytical (Fourier rebinning) type backprojection methods provide fast and reliable reconstruction for the standard X–ray PET model without scatter. Parameters can be chosen in an optimal way. Images can be guaranteed to deliver a certain precision, which depends on scanner geometry and measurement accuracy. Deviations from the model need to be handled in pre- or postprocessing and cannot be included in the image reconstruction algorithm. Also, *a priori* knowledge about activity distribution functions (like positivity) cannot be handled by the algorithms internally.

Iterative (EM, OSEM) type methods provide much more flexibility with respect to changes of the physical and the noise model. Typically, since they can handle the correct noise model, deal with the pure data without rebinnning (list mode) and take *a priori* assumptions into account, they deliver improved image quality. However, that comes with a price tag: fully 3D list mode iterative EM algorithms are typically slow, optimal stopping criteria are not available, and the derivation of exact error bounds is not possible.

References

[1] J. E. Bresenham. Algorithm for computer control of a digital plotter. *IBM System Journal*, 4(1):25–30, 1965.

[2] T. M. Buzug, J. Borgert, T. Knopp, S. Biederer, T. F. Sattel, M. Erbe, and K. Lüdtke-Buzug (Eds.). *Magnetic nanoparticles: Particle science, imaging technology, and clinical applications*. World Scientific Publishing Company, 2010.

[3] M. Defrise, P. E. Kinahan, D. W. Townsend, C. Michel, M. Sibomana, and D. F. Newport. Exact and approximate rebinning algorithms for 3-D PET data. *IEEE Transactions on Medical Imaging*, 16(2):145, 1997.

[4] M. Defrise and X. Liu. A fast rebinning algorithm for 3D positron emission tomography using John's equation. *Inverse Problems*, 15:1047, 1999.

[5] A. P. Dempster, N. M. Laird, and D. B. Rubin. Maximum Likelihood from Incomplete Data via the EM Algorithm. *Journal of the Royal Statistical Society, Series B*, 39(1):1–38, 1977.

[6] K. Fourmont. Non-equispaced fast Fourier transforms with applications to tomography. *Journal of Fourier Analysis and Applications*, 9(5):431–450, 2003.

[7] H. M. Hudson and R. S. Larkin. Accelerated image reconstruction using ordered subsets of projection data. *IEEE Transactions on Medical Imaging*, 13(4):601–609, 1994.

[8] A. C. Kak and M. Slaney. *Principles of computerized tomographic imaging*. Classics in Applied Mathematics. 33. Philadelphia, PA: SIAM, 2001.

[9] F. Natterer and E. L. Ritman. Past and future directions in X-ray computed tomography. *International Journal of Imaging Systems & Technology*, 12(4):175–187, 2002.

[10] F. Natterer. *The mathematics of computerized tomography*. Classics in Applied Mathematics. 32. Philadelphia, PA: SIAM. xvii, p.222, 2001.

[11] F. Natterer and F. Wübbeling. *Mathematical methods in image reconstruction*. SIAM Monographs on Mathematical Modeling and Computation. 5. Philadelphia, PA: SIAM. xii, p.216, 2001.

[12] L. Parra and H. H. Barrett. List-mode likelihood: EM algorithm and image quality estimation demonstrated on 2-D PET. *IEEE Transactions on Medical Imaging*, 17(2):228–235, 1998.

[13] W. H. Press. Discrete Radon transform has an exact, fast inverse and generalizes to operations other than sums along lines. *Proceedings of the National Academy of Science USA*, 103(51):19249–19254, 2006.

[14] A. J. Reader, S. Ally, F. Bakatselos, R. Manavaki, R. J. Walledge, A. P. Jeavons, P. J. Julyan, S. Zhao, D. L. Hastings, and J. Zweit. One-pass list-mode EM algorithm for high-resolution 3-D PET image reconstruction into large arrays. *IEEE Transactions on Nuclear Science*, 49(3):693–699, 2002.

[15] B. Setzepfand. *ESNM: Ein rauschunterdrückendes EM-Verfahren für die Emissionstomographie*. PhD thesis, Dissertation, Fachbereich Mathematik, Universität Münster, 1992.

[16] L. A. Shepp and Y. Vardi. Maximum likelihood reconstruction for emission tomography. *Medical Imaging, IEEE Transactions on*, 1(2):113–122, 1982.

[17] R. L. Siddon. Fast calculation of the exact radiological path for three-dimensional CT array. *Medical Physis*, 12:252–255, 1985.

Part II

Correction Techniques in PET and SPECT

Chapter 4

Basics of PET and SPECT Imaging

Ralph A. Bundschuh and Sibylle I. Ziegler

Rechts der Isar Hospital, Technical University of Munich, Munich, Germany

4.1	Introduction		51
	4.1.1	Interaction of photons with matter	52
		4.1.1.1 Photoelectric effect	52
		4.1.1.2 Compton scattering	52
	4.1.2	Photon attenuation	54
	4.1.3	Scatter	57
	4.1.4	Variation in detector efficiency, normalization	58
	4.1.5	Dead time effects (loss of count rate) (PET and SPECT)	59
	4.1.6	Partial volume effects (PET and SPECT)	59
		4.1.6.1 Spill out	60
		4.1.6.2 Spill in	60
	4.1.7	Time resolution and randoms (PET only)	61
	4.1.8	Collimator effects—Distance dependent spatial resolution (SPECT only)	62
	4.1.9	Positron range and annihilation (PET only)	63
References			64

4.1 Introduction

In the last years, quantitative imaging for preclinical as well as for clinical use became more and more important, mainly in oncology but also in neurology and cardiology. PET and SPECT are imaging modalities facilitating quantitative *in vivo* measurements. For exact quantification, several effects degrading the acquired data have to be taken into account. Some of these effects, such as detector efficiency, partial volume effects, photon attenuation and scatter, are found in SPECT as well as in PET imaging. In contrast, the collimator-detector response occurs only in SPECT and random coincidences or limitations due to positron range are specific for PET. This chapter describes these effects. Potential compensation methods are discussed in the following chapters.

4.1.1 Interaction of photons with matter

Depending on the photon energy, the probability of interaction types between photons and matter varies. The energy transferred to the electrons is lost to the atoms in the absorbing matter, resulting in ionization or excitation. In the photon energy range used in clinical emission tomography (about 80 to 511 keV), the interactions are mostly with electrons of the atomic shell: Photoelectric processes and Compton scattering. Coherent scattering (Rayleigh scattering) is the interaction of the photon with the whole atom and is more important at energies of less than 50 keV. Pair production or triplet production as well as photonuclear interaction are important only for energies higher than 1 MeV. An example for cross sections for the different interactions in carbon, making a big fraction in organic tissue, depending on the photon energy, is displayed in Figure 4.1.

4.1.1.1 Photoelectric effect

In tissue, the photoelectric effect is more important for low energy SPECT radionuclides (about 100 keV) and is negligible for the annihilation photons in PET (Figure 4.1). In the case of a photoelectric interaction, the photon loses all of its energy. This is possible because the atom to which the electron is bound can absorb some of the momentum and hence momentum is conserved. The photon energy is divided into the kinetic energy of the emitted electron and the binding energy of the electron. Mostly inner shell electrons are involved in photoelectric interaction. The emitted electron leaves a hole behind which is normally filled up with an outer shell electron resulting in emission of characteristic X-ray or Auger electrons. The cross section for the photoelectric effect is given by the following equation:

$$\sigma = \frac{32}{3}\sqrt{2}\pi r_e^2 \alpha^4 Z^5 \left(\frac{m_e c^2}{E_\gamma}\right) \tag{4.1}$$

where r_e is the classical electron radius, α the fine structure constant, E_γ the photon energy before interaction, m_e the electron mass, and c the speed of light.

4.1.1.2 Compton scattering

If the photon interaction takes part with an outer shell electron, the latter cannot be considered as tightly bound to the nucleus. Actually if the energy of the incoming photon is much larger than the binding energy of the electron, the latter can be assumed unbound. Hence the photon energy cannot be transferred in total to the electron due to momentum conservation. In contrast to the photoelectric effect, photons undergoing Compton scattering loose only

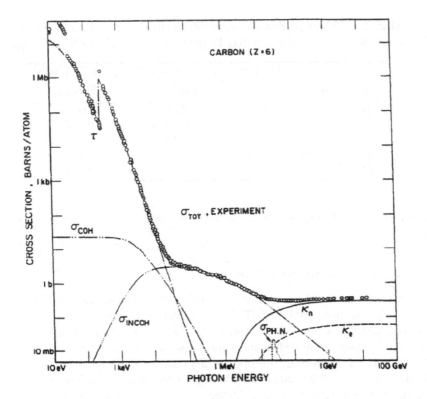

FIGURE 4.1: Different contributions to the total cross section σ_{tot} (circles) in carbon over photon energies ranging from 10 eV to 100 GeV. τ: absorption edge, σ_{COH}: coherent scattering, σ_{INCOH}: Compton scattering, κ_n: nuclear field pair production, κ_e: electron field pair production, and $\sigma_{PH.N.}$: nuclear photo absorption. (From [4].)

some of their energy. The differential cross section for an unbound electron and a photon is given by the Klein–Nishina formula depending on the photon angular distribution:

$$\frac{d\sigma}{d\Omega} = Zr_e^2 \left(\frac{1}{1 + \frac{E_\gamma}{m_e c^2}(1 - \cos\Theta_c)} \right)^2 \left(\frac{1 + \cos^2\Theta_c}{2} \right)$$
$$\left(1 + \frac{\alpha^2(1 - \cos\Theta_c)^2}{(1 + \cos^2\Theta_c)(1 + \frac{E_\gamma}{m_e c^2}(1 - \cos\Theta_c))} \right) \quad (4.2)$$

where Θ_c is the scatter angle, r_e is the classical electron radius, α the fine structure constant, E_γ the photon energy, m_e the electron mass, and c the speed of light. Compton scattering per gram is nearly the same for differ-

TABLE 4.1: Linear photon attenuation coefficients at 140 keV and 511 keV for different materials (from ICRU Report 44 [1] and Hubbell 1969 [3]).

Material	Density [g/cm^3]	μ at 140 keV [1/cm]	μ at 511 keV [1/cm]
Water	1.00	0.150	0.095
Lung	1.05	0.04-0.06	0.025-0.04
Fat tissue	0.95	0.142	0.090
Cortical bone	1.92	0.284	0.178
Muscle	1.05	0.155	0.101

ent materials (Z independent) because electron density is similar for most materials.

4.1.2   Photon attenuation

Photons, in the energy range of SPECT as well as for the 511 keV coincidence photons measured in PET, interact within the patient either via photoelectric effect or Compton scattering. In the case of absorption, the photons will not be detected and their information is lost for the image reconstruction. In the case of scattering, the photon may still be measured in the detector, leading to wrong information for the image reconstruction. This is addressed in the next paragraph (4.1.3). All processes that reduce the number of detected photons are put together under the term *attenuation*. The degree of attenuation of a photon beam depends on the material through which the beam traverses and on the thickness of the material. The attenuation capacity of a material is described by its linear attenuation coefficient μ. Values for some materials can be found in Table 4.1.

Attenuation for a photon beam starting at s' can be described with the following formula:

$$TF(s') = e^{-\int_{s'}^{\infty} \mu(s)ds}. \tag{4.3}$$

While for small animal imaging photon attenuation is often neglected as the effect is quite small as the photons are attenuated by only a few millimeters of tissue, for human PET and SPECT, photon attenuation in the patient leads to a massive underestimation of the activity concentration in PET data, interfering with quantitative and even qualitative image analysis, as discussed later. In Figure 4.2 (top and bottom) effects of photon attenuation in PET can be seen in a cylindrical phantom (20 cm diameter) that was filled with a homogeneous activity concentration. In the center of the phantom the activity concentration seems to be lower by a factor of 10 to 11, compared to the image corrected for attenuation. Besides the false absolute quantification of the radiotracer uptake, the different attenuation in the center and the outer parts of the body can lead to problems in the visual image, too. In patients,

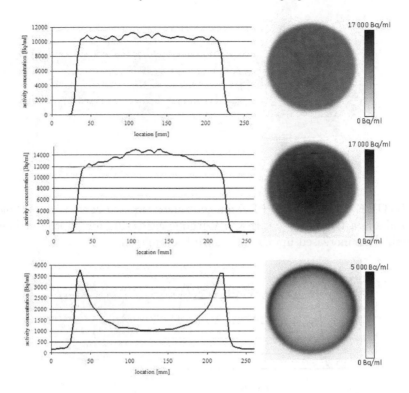

FIGURE 4.2: PET acquisition of a homogenously filled cylindrical phantom with 20 cm diameter. Top image: reconstructed using attenuation and scatter correction; middle image: reconstructed with attenuation correction but without scatter correction; bottom image: reconstructed without corrections. The shown profiles are measured along a vertical line (one pixel wide) through the center of the phantom at the slide that is shown on the right side.

image structures with pathological tracer uptake in a central localization in the body may not be detected due to photon attenuation; an example can be found in Figure 4.3. The influence of photon attenuation for SPECT acquisitions is shown in Figure 4.4 (top and bottom). Also in SPECT the line profile of the image not corrected for photon attenuation shows the same characteristics as for PET; however, the activity in the center of the phantom is underestimated only by a factor of 3. This minor effect is due to the fact that in PET photon attenuation occurs along the whole line of response, whereas in SPECT it occurs only from the center of the phantom to the detector.

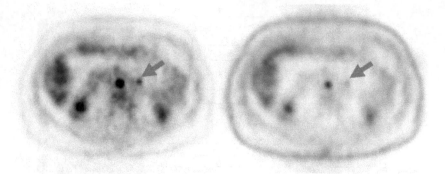

FIGURE 4.3: F¹⁸-FDG PET image (glucose metabolism) with pathological uptake in a lymph node (arrow). Without correction for photon attenuation (right) the increased uptake may have been missed.

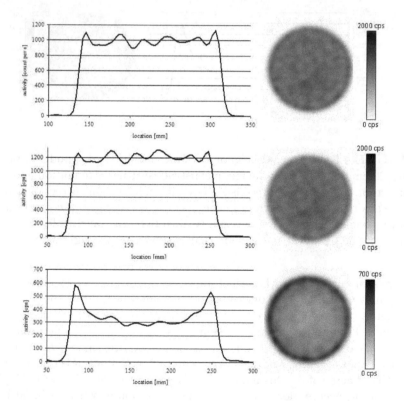

FIGURE 4.4: Tc⁹⁹ᵐ SPECT image of a homogeneously filled 20 cm cylinder phantom. Reconstructed using scatter and attenuation correction (top), attenuation correction without scatter correction (middle), and without any correction (bottom). The profiles at the right side are measured along a horizontal line through the center of the phantom.

4.1.3 Scatter

Scatter does not only lead to a loss of photons for the projection data that would have been counted originally. Also photons that would not have hit the detector on their original path can be detected after scatter as mispositioned events. Photons undergoing Rayleigh scattering do not lose energy; thus, they cannot be distinguished by energy determination. In the energy ranges of SPECT and PET, Compton scattering is the predominant scattering process. In this case, photons loose energy, such that the number of registered scattered events can be reduced by using narrow energy windows in the photon detection process. Scattered events in PET are distributed across the whole field-of-view, accumulating in dense areas, as can be seen in Figure 4.2, where reconstruction was performed with attenuation correction but without scatter correction. In the center of the phantom, the measured activity concentration is up to 15 kBq/ml while the real one is only 11 kBq/ml. The scattered events are attributed more to the center of the phantom because in water, with which the phantom was filled, Compton scatter occurs with a much higher probability than in the air surrounding the phantom. In PET, scattered photons can lead to coincidence detection outside the patient which is illustrated in Figure 4.5 on the left side. In SPECT, scattered events will lead to wrong events only in the patient boundaries (Figure 4.5, right side). Scattered photons that would lead to wrong events outside the patient have no chance to be detected due to the parallel collimation as long as negligible scattering in air is assumed. This can also be seen in Figures 4.3 and 4.4 where for PET the activity concentration in the not-scatter-corrected image (bottom) is not equal to zero

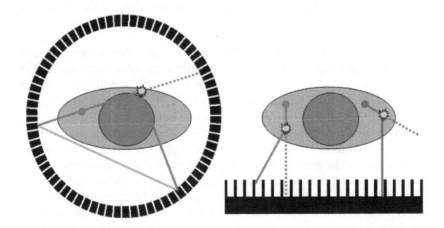

FIGURE 4.5: In PET (left side) coincidence lines can be assigned wrongly to lines outside the patient due to scatter in the body. In SPECT, due to parallel collimation, scattered events can only appear within the boundaries of the body.

outside the image boundaries, while it is for the scatter-corrected case and for the SPECT images (Figure 4.4, bottom).

Besides photon scatter in the object to be measured, scatter can also appear in the scintillation crystals. Due to the higher atomic number of the crystal material even for PET energies, the photoelectric effect plays an important role. Especially in small animal tomographs intra-crystal scatter can lead to relevant reduction of the spatial resolution [6] [8].

4.1.4   Variation in detector efficiency, normalization

PET and SPECT systems do not have a uniform sensitivity for photon detection over their field of view. For SPECT systems there are two typical reasons for such nonuniformities:

1. Nonuniform detection efficiency either due to the fact that the efficiency for detecting light impulses is higher at locations directly over the photomultipliers compared to areas between different photomultiplier tubes or due to differences in the pulse-height spectrum for different photomultipliers.

2. Nonlinearities, describing the effect that straight line objects appear as curved-line images. A well-known example is that when a source is moved toward the center of a photomultiplier tube the light collection efficiency of the photomultiplier increases more quickly than the distance the source is moved. Hence the image of the line source is distorted towards the center of the photomultiplier tube. This typically causes hot spots in the planar image at the location of the photomultiplier tubes.

Other reasons for nonuniformities can be collimator defects, crystal cracking, nonuniformities in the light guide, drop-out of a photomultiplier tube, or defects in the electronics. In contrast to planar scintigraphy, where small nonuniformities may be acceptable, in SPECT nonuniformities can lead to major artifacts, often with the shape of rings. The effect of nonuniformities in peripheral areas of the detector is distributed over large areas in the images, whereas nonuniformities close to the center of the rotation leads to strong artifacts close to the center of the image: the intensity of the artifact is inversely proportional to the distance between nonuniformity and center of the rotation of the detectors. Actual PET scanners, using the block-detector concept, consist of several thousands of detector elements. Each of these detector elements can have its own photon detection efficiency due to inhomogeneities in the detector block. Besides differences between scintillation crystals there can be differences in the efficiency of the light guide and of the photomultiplier tubes as described for SPECT systems before. Additionally in the axial direction, PET systems are built using three to four rings of detector blocks. Between the rings there are gaps without the ability to detect incoming photons. If

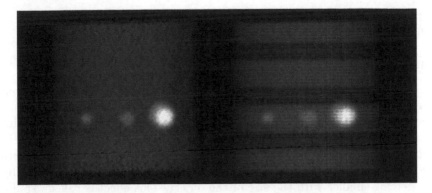

FIGURE 4.6: Phantom with active spheres (coronal slice), reconstructed using normalization file (left) and without correction for normalization effects (right).

no normalization for the image reconstruction is used, these gaps are clearly visible in the reconstructed PET images (Figure 4.6).

4.1.5 Dead time effects (loss of count rate) (PET and SPECT)

Dead time is the time a detection system needs after the detection of one event to be prepared for the detection of another event. This means, if another event occurs in the dead time this new event is ignored. Instead of not detecting the second event, for energy-sensitive detectors the signal of the first event and the second event may overlap and lead to wrong signals. Such merged events may fall outside the energy window of the detection electronics and hence be lost for the image reconstruction. Such events are called dead time losses. The number of dead time losses gets smaller as the dead time of the detection system is shorter. Naturally, higher photon flux leads to a higher probability of another photon not being registered. This means that the measured photons are not proportional to the real photon flux at higher count rates. The fractional dead time of a system is defined as:

$$Dt = \frac{\text{Measured Count Rate}}{\text{Count Rate with Ideal Linear Behaving System}}. \quad (4.4)$$

4.1.6 Partial volume effects (PET and SPECT)

Due to the limited spatial resolution of SPECT and PET tomographs, partial volume effects occur when the acquired objects are smaller than a resolution-volume element of the machine. Partial volume effects lead to false measured activity concentration, either too high or too low depending on the activity distribution [2] [7]. Two different partial volume effects can be distinguished: "spill in" and "spill out."

4.1.6.1 Spill out

Usually, an image volume element contains the measured activity concentration. If the object of interest is smaller than such a volume element, the activity of this object is assigned to the total volume element. Hence, although the total measured activity is correct, the measured activity concentration is smaller than in the original structure. Consequently the measured activity concentration and hence the intensity in the reconstructed image of objects is getting smaller with the size of the object. This effect is called "spill out" as the activity of the object is spilled out into the surrounding volume. This effect can be seen in Figure 4.7 which shows the PET image of a phantom with different sized spheres which all contain the same activity concentration but appear in the PET image as smaller spheres containing less activity concentration.

4.1.6.2 Spill in

Accordingly, activity of surrounding structures can be assigned to the object of interest and hence lead to a false high measured activity concentration. This is of special importance if the object of interest has lower activity concentration than the volume around, for example if the blood activity concentration is measured and the surrounding tissue has higher activity than the blood pool. This is done, for example, for kinetic modeling studies. Often the lumen of the left ventricle of the heart is used for the measurement of blood pool activity; however, for some tracers like FDG there can be high tracer uptake in the surrounding myocardium, especially for later time frames. This

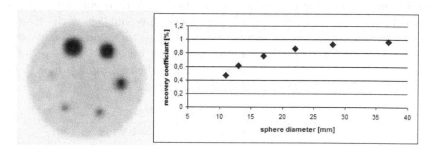

FIGURE 4.7: Partial volume effect: PET image of a phantom filled with ^{18}F-FDG and containing spheres with inner diameters of 37, 28, 22, 17, 13, and 11 mm, respectively. These were filled with ^{18}F-FDG having a 10 times higher activity concentration than the background (left). Corresponding recovery coefficients for the different spheres (right).

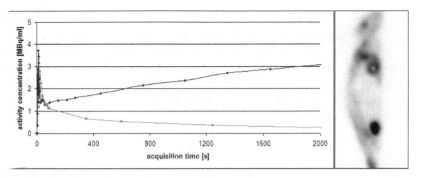

FIGURE 4.8: Blood time activity curve for an F^{18}-FDG mouse study, measured in a volume placed in the left ventricle of the mouse heart (dark gray line). For comparison, the blood activity concentration measured in arterial blood samples is also given (light gray line). Due to spill in from the myocardium the curve measured in the images is still increasing over the time while the real blood activity concentration decreases (left). Corresponding FDG-PET image of a mouse with the VOI (dark gray) in the lumen of the left ventricle where the blood activity concentration was measured (right).

can lead to wrong high values in the time activity curves as shown in Figure 4.8.

A measure for partial volume effects is the recovery coefficient which is defined as:

$$RC = \frac{\text{Measured Activity Concentration}}{\text{Real Activity Concentration}}. \qquad (4.5)$$

The recovery coefficient is dependent on the size of a structure, the activity concentration inside the structure and the surrounding area and even on the geometrical shape of the object. The size dependence of recovery coefficients of spheres filled with an activity concentration 10 times higher than the surrounding background is shown in Figure 4.7 (right side). The recovery coefficient is dependent not only on the structure of interest but also on the reconstruction algorithm of the images and potential post-reconstruction filtering.

4.1.7 Time resolution and randoms (PET only)

In PET, events are measured as coincidences; the two coincidence photons that originate from one positron decay are called true coincidences or trues (light gray event, Figure 4.9). Due to properties of the detector material and the electronics the time resolution of a PET scanner is limited and hence the timing window in which the electronics consider two measured photons as a coincident event cannot be made arbitrarily small. Typical coincidence windows of actual state-of-the-art PET tomographs are around 4 ns. Therefore it can happen that two photons that do not originate from the same annihilation process are measured as a coincidence event (dark gray event, Figure 4.9). These events are called random events or randoms. The rate of random coincidence events along one line of response (LOR) is dependent on the rate of single events for each detector of the LOR (r_i and r_j) and on the length of the timing window (τ):

$$R_{ij} = 2\tau r_i r_j. \tag{4.6}$$

The overall random rate proportionally depends on the square of the activity concentration for a given activity distribution in the field of view. Usually, randoms are distributed homogeneously over the field of view.

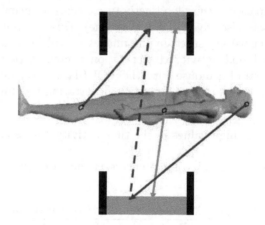

FIGURE 4.9: If photons of two independent annihilation processes are measured in two detectors (dark gray arrows) within the coincidence window of the PET machine they will be considered as a coincidence event. An event along a line of response which is not correlated to the two annihilation processes will be used for image reconstruction (dashed line). Such wrong coincidence events are called random events or randoms. The light gray arrows represent a true coincidence event.

4.1.8   Collimator effects—Distance dependent spatial resolution (SPECT only)

In SPECT collimators are used in front of the scintillation crystals for localization of incoming photons. In clinical routine parallel-hole collimators of different materials, thickness and hole sizes depending on the application are in use. For special applications such as high resolution small animal imaging or cardiac imaging, other collimator types such as converging, diverging, and pinhole collimators are in use. The spatial resolution of a SPECT system depends on several factors: intrinsic spatial resolution of the detector system and the electronics, angular and linear sampling intervals, and the resolution of the collimator. The latter is the dominant factor in most cases. The collimator resolution depends on the geometry of the collimator, mainly the length of the holes as well as their diameter. Besides the collimator characteristics, the spatial resolution of the collimator depends on the distance between collimator and radiation source. For a point or line source this dependency can be estimated by the following equation:

$$r_{coll,spatial} \approx \frac{d(l_{eff} + \Delta_{source})}{l_{eff}} \qquad (4.7)$$

where d is the diameter of the collimator holes, l_{eff} their effective length and $\Delta_{sources}$ the distance between collimator and radiation source. The effective length of the collimator holes is the length of the collimator holes reduced by a correction term taking into account penetration of photons

$$l_{eff} = l - \frac{2}{\mu} \qquad (4.8)$$

with the linear attenuation coefficient μ of the collimator material for the appropriate photon energy.

Besides the effect for spatial resolution, the collimator reduces the number of events measured in the detector, as photons are absorbed, attenuated or scattered in the collimator material. The ratio of events that are detected to the events that fall onto the collimator is called the collimator efficiency. This collimator efficiency is only several percent in typical clinical settings. It is improved if the collimator thickness is smaller (shorter holes) and if the hole diameter is bigger. This means that improving the collimator efficiency always means decreasing the collimator resolution, as a bigger diameter d and a shorter whole length l leads to an increased resolution according to Equation 4.7.

4.1.9 Positron range and annihilation (PET only)

In PET the position of the annihilation process of the positron with an electron is measured. However, the localization of the annihilation is not equivalent to the place of the positron emission from the radiotracer, which is the location of interest. The reason for this is the excess beta decay energy which manifests itself as kinetic energy of both positron and neutrino. As this decay energy can be distributed between the neutrino and the positron differently, the emitted positron has no specific kinetic energy, but an energy spectrum with mean and maximum kinetic energy; the latter can be found in Table 4.2. The annihilation process finally takes place when the kinetic energy is transferred to the surrounding matter. Due to the energy spectrum, the range of the positrons is also not predetermined, but can be described by a distribu-

TABLE 4.2: Maximum positron energy and mean and maximum positron range in water for some typical PET isotopes (adapted from [5]).

Isotope	E_{max} [MeV]	R_{max} [mm]	R_{mean} [mm]
^{11}C	0.96	4.1	1.1
^{13}N	1.20	5.1	1.5
^{15}O	1.70	7.3	2.5
^{18}F	0.63	2.4	0.6
^{68}Ga	1.90	8.2	2.9
^{82}Rb	3.40	14.1	5.9

tion with mean and maximum range in body tissue (water equivalent). These values can be found in Table 4.2 for the most important isotopes in PET.

Another limitation of the spatial resolution of a PET system originating from the annihilation process itself is that the 180-degree angle between the two annihilation photons is valid only when positron and electron have zero momentum at the time of annihilation. Actually, in more than 65% of the annihilation processes this is not true. The angular distribution of the annihilation photons was found by Levin and Hoffman to be Gaussian shaped with 0.25 degree FWHM [5]. The effect on the spatial resolution of the PET system due to this non-collinearity depends on the diameter of the detector ring and is about 1.8 mm for a 80 cm ring system.

References

[1] ICRU Report 44: Tissue substitutes in radiation dosimetry and measurement. Technical report, 1989.

[2] E. J. Hoffman, S. C. Huang, and M. E. Phelps. Quantitation in positron emission computed tomography: 1. Effect of object size. *Journal of Computer Assisted Tomography*, 3(3):299, 1979.

[3] J. H. Hubbell. *Photon cross sections, attenuation coefficients, and energy absorption coefficients from 10 keV to 100 GeV*. US National Bureau of Standards; for sale by the Supt. of Docs., US Govt. Print. Off., [Washington], 1969.

[4] J. H. Hubbell. Review of photon interaction cross section data in the medical and biological context. *Physics in Medicine and Biology*, 44:R1, 1999.

[5] C. S. Levin and E. J. Hoffman. Calculation of positron range and its effect on the fundamental limit of positron emission tomography system spatial resolution. *Physics in Medicine and Biology*, 44:781–799, 1999.

[6] M. Rafecas, G. Böning, B. J. Pichler, E. Lorenz, M. Schwaiger, and S. I. Ziegler. Inter-crystal scatter in a dual layer, high resolution LSO-APD positron emission tomograph. *Physics in Medicine and Biology*, 48(7):781–799, 2003.

[7] M. Soret, S. L. Bacharach, and I. Buvat. Partial-volume effect in PET tumor imaging. *Journal of Nuclear Medicine*, 48(6):932, 2007.

[8] I. Torres-Espallardo, M. Rafecas, V. Spanoudaki, D. P. McElroy, and S. I. Ziegler. Effect of inter-crystal scatter on estimation methods. *Physics in Medicine and Biology*, 53:2391–2411, 2008.

Chapter 5

Corrections for Physical Factors

Florian Büther
*Department of Nuclear Medicine and European Institute of Molecular Imaging,
University of Münster, Münster, Germany*

5.1	Introduction	67
5.2	Decay correction	69
5.3	Randoms correction	71
	5.3.1 Singles-based correction	71
	5.3.2 Delayed window correction	72
5.4	Attenuation correction	73
	5.4.1 Stand-alone emission tomography systems	77
	5.4.2 PET/CT and SPECT/CT systems	80
	5.4.3 Attenuation correction artifacts	82
5.5	Scatter correction	90
	5.5.1 Energy windowing methods	91
	5.5.2 Analytical methods	92
	5.5.3 Direct calculation methods	94
	5.5.4 Iterative reconstruction methods	95
5.6	Concluding remarks	95
References	....	95

5.1 Introduction

The physical basis of both emission tomography technologies, PET
(positron emission tomography) and SPECT (single photon emission computed tomography), is the fact that many atomic nuclei are not stable; they
decay into other nuclei under the emission of specific radiation (*radioactivity*),
a property that was first discovered by Henri Becquerel in 1896 while investigating the fluorescence of uranium compounds. In emission tomography, two
fundamental decay modes are of particular interest: the beta$^+$ decay for PET
and the gamma decay for SPECT.

An atomic nucleus undergoing a beta$^+$ decay emits an electron neutrino
ν_e and a positron e^+, thereby decreasing its atomic number Z by 1 while

retaining its mass number A. One such example is the decay of the often-used PET nuclide ^{18}F:

$$^{18}_{9}\text{F} \rightarrow {}^{18}_{8}\text{O} + e^{+} + \nu_{e}. \tag{5.1}$$

In PET, this decay takes place inside the body of a patient, and the emitted positron loses its kinetic energy by scattering. Once slow enough (after a maximum range in the order of a few mm, depending on the initial kinetic energy of the positron and the nature of the the traversed material), it annihilates with one of the ubiquitous electrons, resulting in most cases in two gamma photons γ of the specific energy of $\text{E}_{\gamma} = 511$ keV:

$$e^{+} + e^{-} \rightarrow 2\gamma. \tag{5.2}$$

These photons are emitted in opposite directions under an angle of 180°. As gamma radiation penetrates soft tissue quite well, it is possible that both photons can be detected almost simultaneously outside the body by gamma-sensitive radiation detectors (coincidence event), giving spatial information about the location of the decay.

The gamma decay does not change the identity of the nucleus; rather, gamma-decaying nuclei are in an energetically excited state that is usually a consequence of a prior alpha or beta decay. The excess energy can then be delivered either in total or partially by emission of gamma radiation of characteristic energies. An example is the gamma decay of the metastable SPECT nuclide $^{99\text{m}}$Tc which decays by emitting 143 keV gamma radiation:

$$^{99\text{m}}_{43}\text{Tc} \rightarrow {}^{99}_{43}\text{Tc} + \gamma. \tag{5.3}$$

If delivered inside the human body, this radiation can be detected outside by detectors. As no coincident events as in PET are measured in SPECT, collimator techniques have to be applied in order to gain information about the decay location.

The detected radiation events during a certain time interval then allow the reconstruction of image data comprising information about the activity distribution within the human body. Both PET and SPECT potentially have the ability to absolutely quantify the dispersed radioactivity *in vivo*. However, the acquired raw data has to be processed either before, during, or after the reconstruction process to get satisfying images in terms of quality and quantity. These corrections can be divided into two groups:

- Corrections for factors related to physical processes during radioactive decay and during the steps until the radiation leaves the body;

- Corrections for factors related to the detection of the radiation.

The first factors will be the topic of the present chapter; the second factors will be discussed in the next chapter.

Ideally, an emission tomography scanner should be able to address the following problems related to the physics of radioactive decay and the interaction between radiation and tissue:

- Physical decay of the radionuclide;

- Loss of gamma photons by attenuation processes while traversing the human body;

- Detection of gamma photons that changed direction due to scattering;

- Detection of coincident gamma photons that do not originate from the same annihilation event (random coincidences).

Decay, attenuation and scatter affect both PET and SPECT imaging, while random coincidences are a specifically PET-related problem. Figure 5.1 shows the different types of events that are connected to these physical effects and which can be measured in PET. (Not shown are multiple events which are usually discarded and not taken into account.) In order to get absolute quantifiable emission tomography data, one has to deal with all these effects to minimize their influence in the reconstructed images. Different correction methods have been investigated since the introduction of emission tomography techniques; the most important ones along with some interesting recent development will be discussed in the following sections. As an exhaustive description of correction techniques in both PET and SPECT (especially considering attenuation and scatter correction) would go beyond the scope of this chapter, the main focus will be put on PET correction techniques.

5.2 Decay correction

The most straightforward correction method in emission tomography is the correction for the physical decay of the employed radionuclide. This is especially necessary in dynamical studies where activity distribution values at different points in time are to be measured. The main task then is to calculate weighting factors that transform the measured activity values for every time frame to values that would have been measured if the activity would have remained constant in time.

A given radioactive decay process is governed by an exponential law:

$$A(t) = A_0 \cdot e^{-\lambda t}, \tag{5.4}$$

where $A(t)$ denotes the activity of a radioactive sample at time t, A_0 is the activity at time $t = 0$, and λ denotes the decay constant of the decay process which is given by $\lambda = \ln(2)/T_{1/2}$ ($T_{1/2}$: half-life of the nuclide). A correction

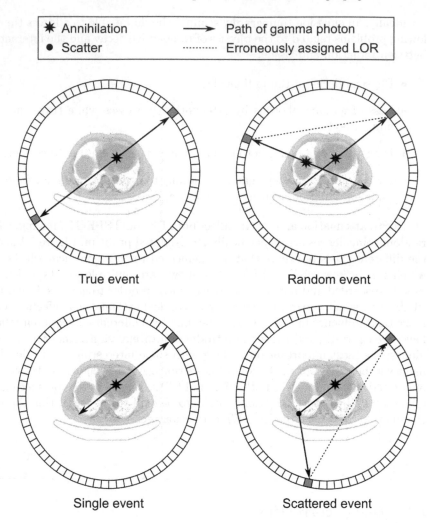

FIGURE 5.1: Measurable events in PET.

factor for an emission event measured during a scan would therefore be $e^{\lambda t}$. For a whole time frame between t_1 and $t_1 + \Delta t$, one has to compare the area under the curve $A(t) = A_0 \cdot e^{-\lambda t}$ from t_1 to $t_1 + \Delta t$ to the rectangular area defined by A_0, t_1 and $t_1 + \Delta t$ (see Figure 5.2). This gives the decay correction factor DCF as

$$DCF = \frac{A_0 \Delta t}{\int_{t_1}^{t_1+\Delta t} A(t)\mathrm{d}t} = \frac{\lambda \Delta t}{1 - e^{-\lambda \Delta t}} e^{\lambda t_1}. \tag{5.5}$$

Generally, as decay correction does not depend on the spatial location of emission events, this correction factor does not contribute to the quality of an

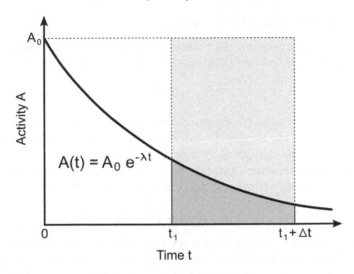

FIGURE 5.2: Radioactive decay. The activity of a radioactive sample as a function of time gives an exponential relationship determined by the nuclide-specific decay constant λ.

emission tomography image; however, it obviously affects the quantitation of image-derived values.

5.3   Randoms correction

The correction for random coincidence events is a PET-specific problem. Random events contribute to both bias in measured activity distribution and a loss of image contrast by adding additional background to the data. The impact of randoms on the acquired images is dependent on several factors such as the length of the coincidence timing window, the administered activity, the energy window limits, and the scanning mode (2D vs. 3D PET).

There are basically two standard strategies that enable estimating of the number of random coincidences between two detector elements. The first one is a statistical calculation of random events based on measured single events; the second one relies on a measurement of random coincidence events with the help of a delayed coincidence window.

5.3.1   Singles-based correction

This approach requires the knowledge of the singles rate on each detector. A statistical consideration then leads to the following relation:

$$R_{ij} = 2\tau_{coin} s_i s_j, \tag{5.6}$$

where s_i, s_j are the single counting rates measured on detector i and j, respectively, R_{ij} is the expected randoms rate between the detectors, and τ_{coin} is the length of the coincidence window [48]. This rate can be subtracted from the measured coincidence rate P_{ij} between the two detectors, resulting in a good estimate of the true coincidence rate T_{ij} (in addition to scattered events S_{ij}):

$$T_{ij} + S_{ij} \approx P_{ij} - R_{ij}. \tag{5.7}$$

An advantage of this method is the low variance of the calculated randoms rate as single rates in PET are much higher than coincidence rates. However, a proper knowledge of system dead times, variances in τ_{coin} between different detector elements and detector efficiencies is crucial for correct randoms estimation, as otherwise the calculated rate may be biased [14].

5.3.2   Delayed window correction

Another method, the delayed window approach, is based on a more direct measurement of random events. For this, a measured count on a detector not only opens the usual prompt coincidence window of width τ_{coin}, but also another window after some time $\tau_{delay} > \tau_{coin}$ (see Figure 5.3). This ensures that delayed coincidences cannot be caused by a single annihilation event and must therefore be random coincidences. Subtracting the delayed coincidence rate D_{ij} from a line of response between detectors i and j from the prompt

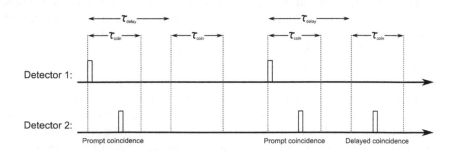

FIGURE 5.3: Estimation of random coincidences using the delayed window approach. Counts on a detector open two windows, the coincidence window and the delayed window. Measured coincidences during the coincidence window can be either true, scattered or random coincidences; coincidences in the delayed window can only be random coincidences.

coincidence rate P_{ij} leads to an estimate of the true coincidence rate T_{ij} (in addition to the scatter rate S_{ij}):

$$T_{ij} + S_{ij} \approx P_{ij} - D_{ij}. \tag{5.8}$$

Dead time effects are of no concern in this approach as both the prompt and the delayed windows have the same dead time properties; however, as the measured delayed rate is much lower than the singles rate, the calculated randoms rate for a given line of response has much more variance as compared to the singles-based method. This can be reduced by storing the collected delayed events in a separate sinogram data set (besides the prompt sinogram data set) instead of an online subtraction of delayed coincidences. The delayed sinograms may then be processed by filtering steps (smoothing) to reduce the variance before subtracting them from the prompt data [20]. This obviously comes at the cost of computer memory and processing time, but helps to reduce the signal-to-noise ratio in the reconstructed images as compared to the simple online subtraction method [14].

5.4 Attenuation correction

A beam of monochromatic gamma rays that is passing through matter is attenuated in intensity. Empirically, the intensity I after passing a homogeneous material of thickness r is found to be

$$I(r) = I_0 \cdot e^{-\mu r}, \tag{5.9}$$

where I_0 is the incident intensity of the beam and μ is the total linear attenuation coefficient, a gamma energy- and material-dependent constant: $\mu = \mu(E_\gamma, \text{material})$. In the case of an inhomogeneous material with spatially variable linear attenuation coefficients $\mu(r)$ (the so-called attenuation map or μ-map), the transmitted intensity is

$$I(r) = I_0 \cdot e^{-\int_i \mu(r) \mathrm{d}r}, \tag{5.10}$$

with i denoting the line of the gamma ray.

This attenuation of intensity is due to interaction processes between the photons and the atoms of the material. Basically, gamma photons discharged in PET and SPECT can interact with matter by two processes known as the Compton effect and the photoelectric effect. (A third process, electron-positron pair production, is possible only at gamma energies above 1.022 MeV which corresponds to twice the rest mass of electrons.)

Compton scattering is an inelastic scattering effect of photons interacting with electrons, most probably those on the outer shell of atoms. The photon transfers energy during the process to an electron and changes its traveling

TABLE 5.1: Approximate μ-values for 140 keV (^{99m}Tc SPECT) and 511 keV (PET) in cm^{-1} (calculated using values from [38])

Material	μ at 140 keV	μ at 511 keV
Adipose tissue	0.143	0.090
Water	0.155	0.096
Soft tissue	0.157	0.098
Cortical bone	0.285	0.172

direction. Connected to the loss of energy is a decrease in wavelength of the gamma photon.

A gamma photon may also interact with a (predominantly inner-shell) electron by transferring all its energy, thereby ceasing to exist and ejecting the electron from the atom. This process is known as the photoelectric effect.

Both effects contribute to the attenuation of gamma rays in matter, and the linear attenuation coefficient can therefore be written as

$$\mu(E_\gamma, \text{material}) = \mu_{\text{Compton}}(E_\gamma, \text{material}) + \mu_{\text{Photo}}(E_\gamma, \text{material}). \quad (5.11)$$

For a given attenuation material, both contributions generally decrease when the gamma energy increases. In PET and SPECT scans which use gamma energies above 100 keV, the contribution of the photoelectric effect to the total attenuation coefficient is negligible for both soft tissue and bone structures, and by far the predominant effect is Compton scattering. This is no longer the case in high-Z material such as lead, where the photoelectric effect is predominant for SPECT energies and roughly equal to the Compton effect at PET energies. Approximate μ-values in PET and SPECT imaging are given in Table 5.1.

In PET, attenuation correction factors are independent of the location of the annihilation event on the line of response because of the fact that the whole line of response is traversed by either one of the gamma photons. This is readily visible in Figure 5.4. If P_{detect} denotes the total detection probability of the photon pair which is created in r_1 on the line of response defined by detectors d_1 and d_2, then obviously

$$P_{detect} \propto e^{-\int_{d_1}^{r_1} \mu(r)dr} \cdot e^{-\int_{r_1}^{d_2} \mu(r)dr} = e^{-\int_{d_1}^{d_2} \mu(r)dr} \quad (5.12)$$

is independent of the specific location of r_1 on the line connecting d_1 and d_2. Therefore, for every line of response in PET, just one attenuation correction factor has to be known to perform attenuation correction which is thus simply done by multiplying the measured number of counts g_i on the line of response i by the attenuation correction factor for this line:

$$g_i^{\text{AC}} = e^{-\int_i \mu(r)dr} \cdot g_i. \quad (5.13)$$

This is not the case in SPECT, where the attenuation correction depends on the location of the point of origin along the detection line and therefore

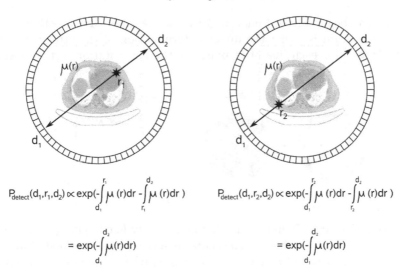

$$P_{detect}(d_1, r_1, d_2) \propto \exp(-\int_{d_1}^{r_1} \mu(r)dr - \int_{r_1}^{d_2} \mu(r)dr) \qquad P_{detect}(d_1, r_2, d_2) \propto \exp(-\int_{d_1}^{r_2} \mu(r)dr - \int_{r_2}^{d_2} \mu(r)dr)$$

$$= \exp(-\int_{d_1}^{d_2} \mu(r)dr) \qquad\qquad\qquad = \exp(-\int_{d_1}^{d_2} \mu(r)dr)$$

FIGURE 5.4: In PET, the detection probability P_{detect} of an annihilation on a given line of response between two detectors d_1 and d_2 is independent of the actual location r_1 (left), r_2 (right) of the annihilation on the line.

makes proper attenuation correction in SPECT a less straightforward task than in PET. One method of correcting SPECT data for uniform attenuation characterized by μ that has found its way into clinical practice is known as the Chang method [21]. It is a multiplicative post-reconstruction correction performed on SPECT images. If $f(x, y)$ and $f^{AC}(x, y)$ denote the non-corrected and corrected activity distribution in (x, y), then according to Chang

$$f^{AC}(x, y) = \frac{f(x, y)}{M^{-1} \sum_{i=1}^{M} \exp(-\mu s_i)} \qquad (5.14)$$

where M is the total number of projection angles and s_i is the distance from (x, y) to the border of the attenuation object in the ith projection. This formula has also been applied in non-uniform attenuation correction, especially in an iterative approach [44].

Instead of pre-reconstruction (PET) or post-reconstruction (SPECT) attenuation correction approaches, this correction step can also be performed during reconstruction using iterative algorithms as the maximum likelihood expectation maximization (MLEM) or the ordered subsets expectation maximization (OSEM) algorithm that easily allow modeling of the attenuation process in the reconstruction. See Chapter 3 for detailed informations on these approaches.

The basic effect of attenuation in emission tomography images is the underestimation of tracer concentrations in regions close to and especially deep inside attenuating objects. Examples for PET are given in Figure 5.5 and Figure 5.6 for a cylinder phantom scan and an FDG patient scan, respectively.

As can be seen from Equation 5.13, the main task in correcting PET data for attenuation is either the direct determination of the attenuation correction factors for each line of response or the computation of the attenuation

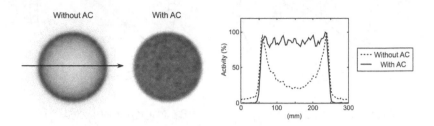

FIGURE 5.5: PET image of a homogeneously-filled cylinder phantom containing ^{68}Ge. The activity in the center is severely underestimated in the non-attenuation-corrected image (left) as compared to the properly attenuation- and scatter-corrected image (middle) as demonstrated by the profile along the indicated arrow (right).

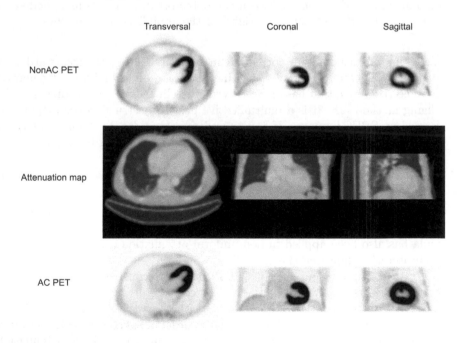

FIGURE 5.6: FDG PET scan of the thorax showing non-attenuation corrected (top) and attenuation-corrected image (bottom) along with the attenuation map $\mu(r)$. Without correction, structures like liver and heart interiors are suppressed in terms of tracer uptake, while uptake in tissue with small μ values, like the lungs, is over accentuated.

map $\mu(r)$ with which the correction factors can then easily be calculated. In the following sections, several approaches that have been applied in PET and SPECT systems are discussed. Although attenuation in emission tomography is a well-understood phenomenon, these correction approaches themselves may introduce new problems leading to image degradation and biased quantification requiring strategies to minimize these problems. This is subject of current research and shall be discussed in yet another section.

Besides these methods that are based on a measurement of transmission data, several methods based on calculating attenuation correction factors have been investigated [92]. However, as these are usually restricted to certain applications in emission tomography, they are not discussed here.

Recent development of hybrid PET/MR systems has led to new challenges in the field of attenuation correction which will be discussed in detail later in this book.

5.4.1 Stand-alone emission tomography systems

In stand-alone PET systems, attenuation correction factors are usually determined by performing an extra transmission scan besides the ordinary emission scan. For this approach, one or more positron-emitting sources, either in ring form or as rod sources [19], located inside the detector ring rotate around the scanned object; see Figure 5.7. These sources usually contain a few mCi of ^{68}Ge which decays into the positron emitter ^{68}Ga with a half-life of 271 days. In order to avoid excessive dead time when using just one rod source [31], two or three rod sources with accordingly reduced activity are usually used,

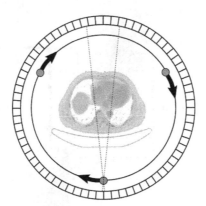

FIGURE 5.7: Transmission scan using three rod sources rotating around the patient. When using sinogram windowing, only coincidence events that pass through one of the sources (for example the dashed lines) are taken into account, therefore reducing the bias introduced by random and scattered events.

thus avoiding excessive radiation events seen by a single detector element. The released annihilation radiation of these sources is registered just as in the emission scan, resulting in transmission sinograms t_i that contain information about the attenuation along all lines of response i. To derive attenuation correction factors, a second positron source-based scan, the so-called blank scan, has to be performed without any object inside the scanner's field of view, resulting in blank sinograms b_i. In its simplest version, the attenuation correction factor for a line of response i is then given by $\frac{b_i}{t_i}$, which leads to the corrected emission data g_i^{AC}:

$$g_i^{\text{AC}} = \frac{b_i}{t_i} \cdot g_i \tag{5.15}$$

(see Figure 5.8). The main advantage of this method as opposed to other approaches where assumptions about attenuation values are made is the direct

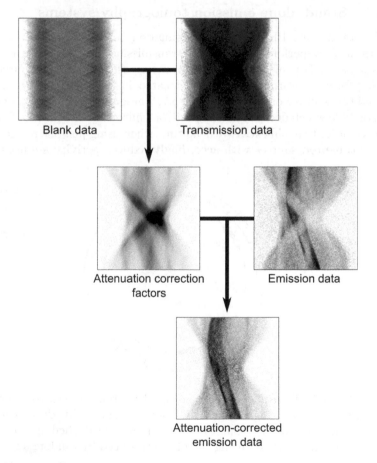

FIGURE 5.8: Measured attenuation correction in stand-alone PET systems.

measurement of the proper attenuation in every scan session. However, this usually comes at the cost of long transmission scan times to ensure a satisfying statistical quality in the transmission data as poor transmission data will lead to an excessive amount of noise in the attenuation-corrected emission data and thus to inferior image quality.

A first step in solving this problem is a spatial smoothing of the blank and transmission data [74]. More advanced methods involve the utilization of *a priori* knowledge of attenuation values μ for different tissue values [65] [87] [88]. For this approach, the noisy attenuation map is first calculated by reconstructing

$$\log \frac{b_i}{t_i} = \int_i \mu(r) \mathrm{d}r, \qquad (5.16)$$

obtaining an image of the attenuation map $\mu(r)$. Then, $\mu(r)$ is segmented using histogram techniques into a limited number of body tissues (soft tissue, lung tissue, bones) whose average μ values are known (see Table 5.1), resulting in a noise-free attenuation coefficient map from which the attenuation correction values can be determined by calculating the line integrals along each line of response. Furthermore, to account for individual variations in μ-values, a weighted sum of both the segmented and the original attenuation map can easily be generated (see Figure 5.9).

Other strategies to avoid long transmission scan times involve operating the blank and transmission scans in singles rather than in coincidence mode as the singles counting rate is much higher than the coincidence counting rate [26]. Obviously, this singles mode can also be applied when using gamma decaying sources like ^{137}Cs ($E_\gamma = 661.7$ keV) instead of positron emitters. This reduces the cost of replacing the sources after a short time, as the half-life of ^{137}Cs is 30 years, ensuring a long system life time [42]. However, if the gamma energy is significantly different from 511 keV, then the measured attenuation data has to be scaled to PET energies, which again can be done in a segmentation approach similar to the one described above [11]. Additionally, the singles-based transmission data is heavily scatter-contaminated as scattered singles events cannot be distinguished from non-scattered events as opposed

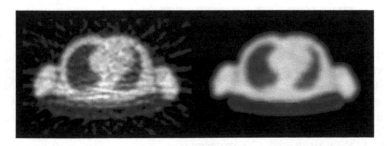

FIGURE 5.9: Attenuation maps derived without segmentation (left) and with segmentation (right).

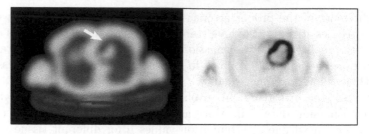

FIGURE 5.10: Determined attenuation map in a cardiac FDG PET scan (left). In this case, post-injection transmission data is heavily contaminated by emission events despite sinogram windowing, resulting in an apparent decrease in attenuation values (arrow) in regions of extremly high myocardial tracer uptake in the emission data (right).

to coincidence-based transmission scans where events that do not traverse the current rod source position can be rejected as scattered coincidences (sinogram windowing [39], Figure 5.7). This generally leads to an underestimation of attenuation and therefore to biased quantification in the reconstructed PET images. Sinogram windowing also has the advantage that transmission scans can be performed after injection of the radiotracer (post-injection transmission scans), therefore reducing total scan time for those radiotracers that need long uptake times like FDG. Without sinogram windowing, post-injection transmission data can be heavily contaminated by emission events. These are mostly suppressed by sinogram windowing. However, in cases of extremely high tissue uptake inside the field of view, emission events may still play a role even in the case of sinogram windowing, as demonstrated in Figure 5.10. In these cases, a subtraction of the estimated emission rate during the transmission scan using data from the emission scan can successfully reduce this bias [19].

For exhaustive reviews of transmission scan techniques in SPECT, the reader is referred to a number of articles where different hardware approaches and their advantages and disadvantages are discussed [44] [92].

5.4.2   PET/CT and SPECT/CT systems

Medical imaging using X-ray-based computed tomography (CT) in combination with emission tomography techniques, especially PET, using a single patient bed system has proven to be a powerful tool to assess both anatomical (CT) and molecular (PET) image data in a single scanning session. The success of combined PET/CT scanner systems since their first availability in clinical practice in 2000 [10] becomes clear when considering that all new clinical PET systems sold nowadays actually are PET/CT systems.

In hybrid PET/CT and SPECT/CT systems, attenuation correction is usually done using the acquired CT image data. CT-based PET attenuation correction offers several advantages over the aforementioned radionuclide-

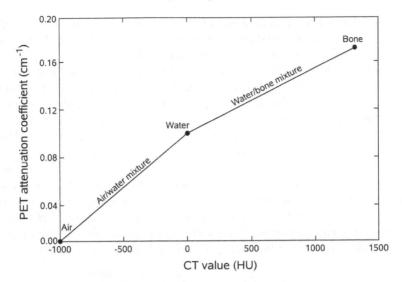

FIGURE 5.11: Bilinear transformation of CT attenuation values given in HU to PET attenuation values. Note that CT values between -1000 and 0 HU are assumed to represent basically a mixture of air and water, while values above 0 HU are considered to be water/bone mixtures.

based transmission scan techniques described above. It is especially valuable in clinical daily routine as attenuation data can be obtained in the order of a few seconds due to fast scanning and reconstruction that is possible in modern CT systems. Due to very high photon flux during CT scans, noise is usually no problem in CT data; besides, this high flux easily allows post-injection transmission scans. Furthermore, X-ray tubes do not decay as radionuclide sources do, thus minimizing replacement costs.

CT images basically are linear attenuation coefficient maps $\mu^{CT}(r)$ of the scanned objects for the utilized X-ray beams. It is important to notice that the used radiation is not monochromatic in nature as in PET (511 keV), but polychromatic. For example, radiation released during a typical CT scan by the X-ray tube that is run at a tube voltage of 130 kV has a continuous spectrum of energies of up to 130 keV Bremsstrahlung in addition to characteristic peaks, altogether with an effective energy of around 70 to 80 keV. Values in CT images are given in Hounsfield units (HU). These are attenuation values that are scaled to the linear attenuation coefficient of water at the effective CT energy:

$$HU = 1000 \left(\frac{\mu^{CT}}{\mu^{CT}_{H_2O}} - 1 \right). \tag{5.17}$$

Thus, water in CT images has 0 HU, while air is at -1000 HU.

The main challenge in PET attenuation correction on PET/CT systems is therefore to transform accurately the CT image data acquired at an effective energy to proper attenuation coefficient values at 511 keV. Unfortunately, knowledge of μ^{CT} alone is not sufficient to determine the proper value of μ^{PET}. This is basically because the contribution of Compton scattering to the attenuation coefficient is depending on the (electron) density of the material alone, while the contribution of the photoelectric effect is additionally strongly dependent on the effective atomic mass Z_{eff} of the material. Therefore, materials having the same μ^{CT} may actually have different μ^{PET} [86]. However, the composition of human bodies usually allows additional constraints leading to a basis for a reliable transformation, as in terms of different attenuation behavior, the human body basically consists of three materials and mixtures thereof: air, water, and bones. Soft tissue with varying density can be well modeled as an accordingly varying mixture of water and air. This simple decomposition into mixtures of air/water and water/bone compartments leads to the following scaling between CT and PET attenuation values

$$\mu^{\mathrm{PET}}(\mathrm{HU} \leq 0) = \mu^{\mathrm{PET}}_{\mathrm{H_2O}} \cdot \frac{\mathrm{HU} + 1000}{1000} \tag{5.18}$$

$$\mu^{\mathrm{PET}}(\mathrm{HU} > 0) = \mu^{\mathrm{PET}}_{\mathrm{H_2O}} + \mu^{\mathrm{PET}}_{\mathrm{H_2O}} \cdot \frac{\mathrm{HU}\left(\mu^{\mathrm{PET}}_{\mathrm{Bone}} - \mu^{\mathrm{PET}}_{\mathrm{H_2O}}\right)}{1000\left(\mu^{\mathrm{CT}}_{\mathrm{Bone}} - \mu^{\mathrm{CT}}_{\mathrm{H_2O}}\right)} \tag{5.19}$$

with predetermined values for $\mu^{\mathrm{PET}}_{\mathrm{H_2O}}$, $\mu^{\mathrm{PET}}_{\mathrm{Bone}}$, $\mu^{\mathrm{CT}}_{\mathrm{H_2O}}$, and $\mu^{\mathrm{CT}}_{\mathrm{Bone}}$ (see Figure 5.11) [15]. It should be mentioned that a change in effective X-ray energy (by changing tube voltage settings) requires an adjustment of these values. Similar transformations have been introduced for scaling CT values to SPECT attenuation values [12] [83]. Alternatives to bilinear transformations have been suggested [43].

Although now often considered a gold standard in clinical practice, CT-based attenuation correction approaches in emission tomography potentially introduce new sources of image artifacts and erroneous tracer quantification. As these problems are important subjects of current research, they are be discussed in detail in the next section.

5.4.3 Attenuation correction artifacts

As already mentioned, the main advantages of CT-based attenuation correction over transmission scan-based attenuation correction using external radionuclide sources are time efficiency (a CT scan can be performed within seconds, while an ordinary transmission scan usually takes at least a few minutes) and low noise properties of the acquired image data due to the much higher photon flux during the CT scan as compared to conventional radionuclide transmission scans. The latter point also allows post-injection transmission scans in PET/CT and SPECT/CT systems without correcting for emission event contamination which can be a problem in radionuclide-based transmission scans as described above.

While these points are very valuable especially in clinical context, there are also some specific disadvantages when using CT data for attenuation correction. One of these problems is based on the aforementioned basic assumption that from an X-ray point of view, human tissue can be divided into mixtures of just three basic materials (air, water, and bone). In some scans, this assumption cannot be maintained because other materials that do not behave like air, water or bone are present. One such instance that is very frequently seen in clinical scans is the usage of contrast-enhanced CT image data for attenuation correction. Shortly before contrast-enhanced CT scans, a contrast agent is given to the patient. This can be done either by intravenous injection into the vascular system to delineate blood vessels from other soft tissue structures or by oral ingestion where the contrast agent is concentrated in the intestine. Chemically, these contrast agents contain atoms of high Z elements like barium and iodine. This ensures an exceedingly high photoelectric absorption, resulting in high attenuation coefficients at X-ray energies and thus good contrast to soft tissue in CT images. Transformation algorithms such as the one described above will treat regions containing a contrast agent as bone-like tissue due to their high X-ray attenuation. However, for these contrast agents, the influence of photoelectric absorption on total attenuation coefficient decreases when going to PET energies in a much stronger manner than a mixture of bone and water. Thus, the determined attenuation coefficients in those regions will be overestimated which in turn should lead to an increase in tracer uptake as compared to attenuation-corrected images using native non-enhanced CT data. Indeed, this type of artifact can be observed in clinical scans (see Figure 5.12). Additionally, the case of bolus injections right before or even during the CT scan raises the question whether transient concentrations of contrast agent in the vascular system will represent a valid attenuation map even if a correct scaling to PET energies can be found.

The question whether contrast agent-induced overestimation in tracer quantification has a significant impact on clinical diagnoses and treatment planning has been the subject of many studies. One study demonstrated PET image artifacts in veins caused by intravenous injection of an iodine-containing agent in 4 out of 30 oncology PET/CT scans [4]. These artifacts were caused by agent concentrations that had significantly higher CT values than in those cases without artifacts. Another study demonstrated significant overestimation of myocardial FDG uptake quantification when using contrast-enhanced CT data for attenuation correction in cardiac PET/CT [17]. However, numerous authors suggest that CT contrast agents intravenously given actually do not result in clinically significant image artifacts or lesion uptake overestimation [8] [64] [89]. This problem seems to be of greater significance when an oral contrast agent is given prior to the CT scan due to a high range of possible concentrations and large distribution volumes [23]. Besides having two CT scans (one without a contrast agent for attenuation correction, another contrast-enhanced one for diagnosis) [84], correction methods addressing this problem have been developed. These mainly include different segmentation

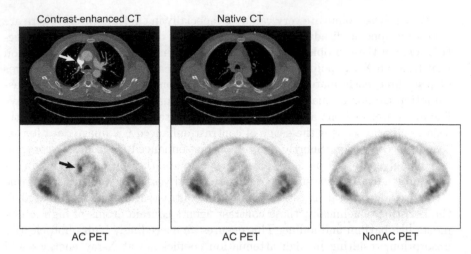

FIGURE 5.12: Contrast agent-induced artifact. The contrast-enhanced CT image (left) shows very high X-ray density in the vena cava (arrow) due to bolus passage of a contrast agent during CT leading to an apparent focal FDG uptake in the attenuation-corrected PET image (arrow). Using a native non-enhanced CT image (middle) for attenuation correction results in no abnormal FDG uptake in this region. The non-attenuation-corrected PET image (right) also demonstrates no abnormal uptake.

steps on the CT image to delineate bones from contrast agent-contaminated tissues. After segmentation, contrast-enhanced regions are properly scaled to PET energies to account for the deviant properties of the contrast agent. The main difficulty here is to find a proper segmentation; investigated techniques comprise segmentation on simple geometrical shapes [17] or manually defined regions [68] as well as more complex ones that apply region-growing algorithms [18] or combined region-boundary-based techniques [2].

Still other methods consider the use of negative oral contrast agents, i.e., substances with a lower CT density than human soft tissue. One such example is water which, when given together with some additives that positively influence its intestinal absorption, was shown to be an effective contrast agent in bowel CT scans [28]. A well-suited agent of this category that has been successfully used in PET/CT is a solution containing locust bean gum and mannitol [5].

Yet another solution would be the use of dual energy CT data for attenuation correction. In this approach, CT data at two different effective energies are acquired. The specific attenuation behavior of contrast agent between the two energies can be used to delineate contrast-enhanced regions from bone material. This was shown to reduce contrast agent-induced image artifacts when compared to single energy-based attenuation correction [78].

Similar problems arise in PET/CT and SPECT/CT scans if metallic im-

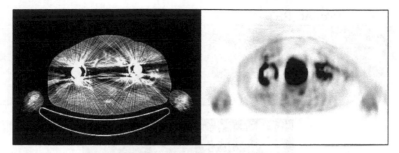

FIGURE 5.13: Typical streak artifacts due to metallic hip prosthetics in CT (left). These artifacts heavily influence quantitation in regions close to the prosthetics in the attenuation-corrected PET image (right).

plants such as hip prosthetics [33], dental implants [41], or implanted pacemaker leads [27] are present. Due to high photoelectric absorption at CT energies in metals, CT images with such implants often show typical streaklike artifacts of both over- and underestimated X-ray density which in turn propagate into the attenuation-corrected emission tomography images leading to difficulties in determining quantitative image data in regions close to the metal (see Figure 5.13). Numerous algorithms have been developed to successfully reduce these CT image artifacts (see [53] and [61] for examples), and this preprocessing of CT data has also been applied to CT data for use in PET attenuation correction [79]. Nevertheless, one study showed that in the case of dental implants, correcting the CT data with one such algorithm did not result in significantly different PET uptake values even in regions affected by this type of artifact [67].

One of the major causes of attenuation correction-induced loss of PET and SPECT image quality is based on the different time scales on which CT and emission tomography techniques operate. While a CT scan can be performed within a few seconds and can thus be taken at breath hold, both PET and SPECT scans usually require several minutes of acquisition time per bed position to achieve satisfying image statistics. Thus emission tomography images usually comprise data of several respiratory cycles, whereas CT images usually represent just a snapshot of a cycle. If the acquired CT data was acquired during a phase that does not match the average PET or SPECT phase, a spatial misalignment between CT and PET/SPECT data will result, which besides hampered image fusion may potentially lead to artifacts due to an erroneous assignment of attenuation correction factors. This problem is of special importance in imaging of lung, liver and heart due to the large gradients in attenuation coefficients between liver/heart and lung tissue, possibly resulting in erroneous tracer quantification or even wrong diagnoses. Clinically, the most frequently seen artifacts of this kind are cold areas at the lung base, indicating that the CT was done in too deep inspiration to match the mean respiratory phase during PET (see Figure 5.14). Several studies have demonstrated the

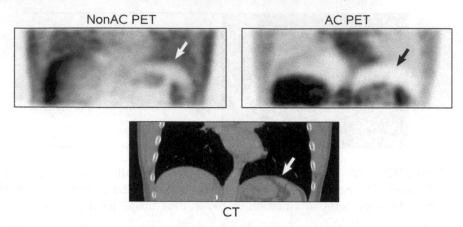

CT

FIGURE 5.14: FDG PET/CT scan showing typical curvilinear cold arti-fact at the lung base (black arrow) due to respiration-induced misalignment between PET and CT (white arrows).

sometimes severe impact of this problem in oncology PET/CT scans [9] [73], especially when using CT data acquired during inspiration (see Figure 5.15). Although first studies in cardiac PET/CT demonstrated accurate attenuation maps generated by CT [49], several other studies have shown that misregistra-tion artifacts in both cardiac PET/CT [36] [52] [63] and SPECT/CT [35] are also quite common, often resulting in false-positive findings especially in the antero-lateral region as demonstrated in Figure 5.16. In stand-alone emssion tomography, this type of artifact is usually not as severe as in hybrid sys-tems, as both emission and transmission scans require several minutes, thus effectively canceling out differences in respiration between them.

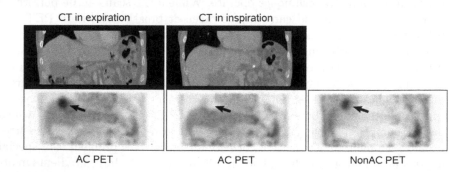

FIGURE 5.15: PET/CT images of a liver tumor close to the diaphragm. While the tumor (arrow) is clearly visible in the non-corrected (right) and the corrected PET image (left) using CT data in end-expiration, it is almost invisible when using end-inspiration CT data for attenuation correction (left).

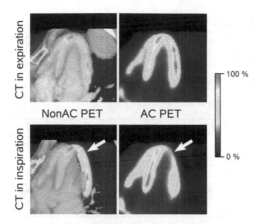

FIGURE 5.16: **(See color insert.)** Cardiac FDG PET/CT scan (long axis view). Non-corrected PET images showing position of the heart during CT (top: end-expiration; bottom: end-inspiration) and PET are shown on the left side, attenuation-corrected PET images on the right. Note the apparent uptake defect in the lateral wall (arrow) due to misregistration between end-expiration CT and PET.

Several approaches to this potential drawback have been discussed in the recent literature. One method proposed in cardiac PET/CT and SPECT/CT is the manual alignment of emission and CT image data [63] [34] which however requires user interaction; additionally, as respiratory motion is usually not confined to pure translational motion, it is not clear to what extent the CT data can be properly aligned to the PET data [36]. Another method would be to try to improve the CT protocol regarding adapted breathing commands for the patient to minimize transmission and emission misregistration [32].

Alternatively, one could aim to simulate the situation in stand-alone systems, i.e., acquiring CT data over one or several respiratory cycles. This can be done either by averaging CT data sets taken at several different respiratory steps in time (Cine CT) [3] [75] or by acquiring a very slow and long CT scan at low tube current, thereby introducing the motion blur into the measured transmission data [71] [82]. Both methods were shown to successfully reduce misalignment between emission and transmission data.

Besides reducing misalignment effects, simulating a stand-alone system obviously introduces motion effects into the transmission data which in the case of large respiratory displacements may degrade image quality due to image blurring. Thus, several methods aim to reduce the impact of respiratory motion in emission tomography data. These approaches can be referred to as motion compensation or motion correction methods. The basic idea here is to use motion information acquired during the scan to divide the raw data into several subsets (gates) of much reduced motion, comprising just one phase of the respiratory cycle. This data can then be used to get near motion-free

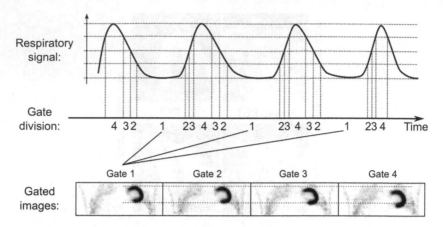

FIGURE 5.17: Respiratory gating in cardiac PET using four gates. Emission data is divided into subsets with reduced motion characteristics according to a respiratory signal acquired during the scan. Shown here is an amplitude-based gating approach. Respiratory motion is clearly resolved in the gated images.

attenuation-corrected image data. The whole process is known as respiratory-gated emission tomography [57] (Figure 5.17). Respiratory gating was shown to be of advantage in terms of uptake quantification in both oncology [70] [60] and cardiac [57] PET scans. Examples demonstrating the gain in tracer quantification and spatial resolution using respiratory-gating are shown in Figure 5.18 for cardiac PET and Figure 5.19 for a PET scan of lung tumors.

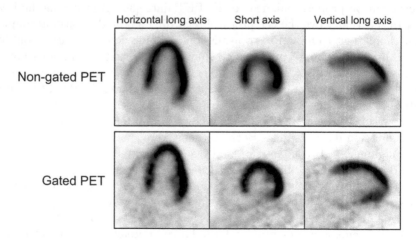

FIGURE 5.18: Cardiac FDG PET/CT scan. The respiratory-gated images demonstrate superior spatial resolution and minimized image artifacts, albeit at the cost of image statistics resulting in higher noise levels.

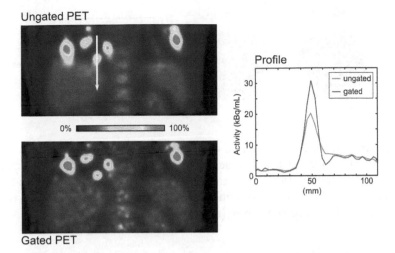

FIGURE 5.19: Non-gated (top) and gated (bottom) FDG PET images of tumors located in the lower lung, coronal view. The arrow denotes the line along which the profile (right) was determined. Respiratory gating reduces motion-induced image blur and enhances tracer quantification.

One of the key elements of respiratory gating is the acquisition of a valid respiratory signal during the scan. For this task, different implementations have been introduced. Commonly used are pressure-sensitive devices fixed at the patient's abdomen during the scan that record changes in pressure due to respiration [46] [51], IR or video camera techniques monitoring markers placed on the patient [25] [69] or devices measuring changes in temperature of the exhaled air [13]. Generally, such a setup should be able to give absolutely quantifiable motion information instead of just giving time information of inspiratory and expiratory phases. This is due to the fact that patients usually do not breathe regularly, neither in terms of frequency nor in terms of amplitude, resulting in a loss of motion resolution when using only time-/phase-based gating. Instead, amplitude-based should be used whenever possible, leading to superior motion capturing [25].

Recently, the idea of using PET raw data (list mode data) itself to obtain respiratory motion information has been discussed, rendering additional hardware unnecessary [16] [80]. This may also have the additional advantage of directly measuring internal organ/lesion motion instead of having only external motion information.

One drawback of respiratory gating is the apparent loss of image statistics, as a significant part of raw data is discarded and not taken into account for reconstruction. However, gating may just be a first step in more generalized motion correction algorithms that aim to reconstruct motion-free images without loss of statistics applying a respiratory motion model during recon-

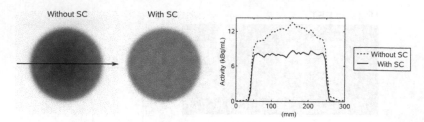

FIGURE 5.20: PET image of a cylinder phantom containing ^{68}Ge. Without correcting for scattered coincidences, the overall activity is overestimated and inhomogeneous (left). Subtracting the scattered events by proper scatter correction (middle) leads to homogeneous activity levels as demonstrated by the line profile along the arrow (right).

struction derived from gated data. Studies in this field have incorporated rigid-body transformations [58], affine transformations [50] and elastic motion models [24]. Detailed insight into these methods are given in Chapter 8.

5.5 Scatter correction

Attenuation correction aims to correct for the loss of photons along a line due to photoelectric absorption or Compton scattering. However, as Compton scattered photons are not lost for detection (in contrast to photons undergoing photoelectric absorption), the acquired raw data are contaminated by these scattered photons. As the photons after scattering no longer possess valid information about their point of origin, they have to be removed (subtracted) from the raw data to ensure absolute quantification of radioactivity (see Figures 5.20 and 5.21). Different strategies for this task have been developed for

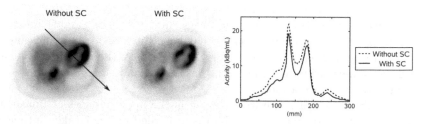

FIGURE 5.21: Transverse PET images of a cardiac FDG patient study. Scatter correction leads to improved image contrast (middle) as evidenced by the line profile along the shown arrow (right).

both SPECT and PET; the most important ones shall be described in this section.

Compton-scattered photons have lost a certain amount of their energy to the electron which is connected to the scatter angle θ by the following formula:

$$E_\gamma^{sc} = \frac{E_\gamma}{1 + \frac{E_\gamma}{m_e c^2}(1 - \cos\theta)}, \tag{5.20}$$

with the electron mass m_e and E^γ and E_γ^{sc} denoting the photon energy before and after scattering, respectively. In principle, scattered events in SPECT and PET should therefore easily be determined by exact measurement of photon energies. Unfortunately, the energy resolution of gamma detectors is by far not good enough to make this a feasible task. This is well demonstrated by current commercial PET detector systems which usually accept gamma photons from an energy range of typically 350 to 650 keV (photopeak window) as annihilation photons. Narrowing this energy window would decrease the amount of scattered events, but would also result in a drastic decrease in overall detection sensitivity.

The overall impact of scatter in emission tomography is generally seen as a loss in image contrast and thus in absolute tracer quantification (see Figure 5.20). The magnitude of this impact is dependent on many factors. Among these are the specific energy window settings, the object that is scanned (and therefore the geometry, the activity distribution, and the attenuation properties of the object) and the geometry of the scanner. Another important issue in PET determining the amount of scatter is the acquisition mode (2D versus 3D): the impact of scattered coincidences is usually much higher in septaless 3D PET scans than in 2D PET scans of the same object. In fact, the amount of scattered coincidences is typically 10–20% of all coincidences (scatter fraction) in 2D studies, while it may increase to more than 50% in 3D PET [91]. Additionally, scatter from outside the field of view is an important contribution to the measured coincidence counting rate in 3D PET and should be corrected for by scatter correction algorithms.

The most important methods in PET scatter correction can be divided into four different groups. They comprise methods based on one of the following ideas: (1) determining scatter by different energy window measurements; (2) calculating scatter by simple analytical models; (3) directly calculating scatter by more complex simulations, either analytically or Monte Carlo-based; and (4) estimating the low frequency scatter distribution during iterative reconstruction of PET data. Each of these ideas will be introduced below.

5.5.1 Energy windowing methods

Although the energy resolution of gamma detectors used in clinical emission tomography systems does not allow a straightforward discrimination between true and scattered events, energy-selective measurements using more than one energy window may give clues to the scatter distribution. These methods have first been developed for use in SPECT systems [40] [45] [30] and have been translated to PET. The main difference between these methods is the way the energy windows are chosen. In the dual energy windows approach for PET [37], the standard photopeak window is set between 380 keV and 850 keV, an additional, lower energy window is set between 200 keV and 380 keV. Both windows will contain scattered and unscattered events (unscattered events that do not deposit their full energy in the detectors will appear in the lower window). The data measured during the scan in both windows is then used in combination with predetermined line source scan data with and without scatter medium to determine the scatter distribution in the standard photopeak window. The true, unscattered events in this window can then be found by subtracting the smoothed scatter distribution from the total event rate. The dual energy window method showed good results in terms of accuracy and ease of implementation in a study comprising numerical simulations, hardware phantom scans, and clinical patient data [90].

The triple energy method makes use of two additional energy windows besides the usual photopeak energy window. These windows are overlapping and have the same upper level setting of 450 keV. The ratio of events in the two windows for the scanned object and for a homogeneous cylinder defines a calibration that can be used to determine the scatter rate in the photopeak window [81]. Other energy windowing methods comprise the usage of multiple energy windows (typically $16 \times 16 = 256$ for coincidence events; multispectral method [7]); however, this mode of acquisition may be limited by the scanner and may thus not be installed easily on PET systems.

Recently, energy-based scatter estimation in PET has been realized by taking detailed energy information available in list mode data into account during image reconstruction [77].

5.5.2   Analytical methods

Analytical methods aim to determine the scatter distribution from the emission data acquired in the photopeak window. This is done using a simple analytical model for the scatter distribution. Two of these approaches based on deconvolution and tail fitting shall be described here in greater detail.

As in the energy windowing methods described above, the acquired raw data g_0 in emission tomography can be described as a sum of both unscattered events g_u and scattered events g_s:

$$g_0 = g_u + g_s. \tag{5.21}$$

In deconvolution-based scatter correction methods, g_s is now modelled as a spatial convolution of the unscattered distribution g_u with a spatially-dependent scatter function f and a scatter fraction k that both describe the scatter properties of the scan:

$$g_s = k \cdot (g_u \otimes f). \tag{5.22}$$

In 2D PET, this convolution is a simple 1D operation on the projection data while in 3D PET it is a 2D convolution on the projection planagrams [6]. The main task now is to solve the equation for g_u. This can be done by direct deconvolution approaches [29] or by convolution-subtraction methods [6]. The former makes use of the convolution theorem of Fourier transformations

$$g_u = F^{-1} \left(\frac{F(g_0)}{F(\delta + k \cdot f)} \right), \tag{5.23}$$

with the Fourier transformation F and the Dirac distribution δ, while the latter uses an iterative approach to determine g_u:

$$g_u^{(1)} = g_0 - k \cdot (g_0 \otimes f) \tag{5.24}$$

$$g_u^{(n)} = g_0 - k \cdot \left(g_u^{(n-1)} \otimes f \right), \tag{5.25}$$

where $g_u^{(n)}$ denotes the nth iterative estimate of g_u. Both scatter fraction k and scatter function f are usually modeled according to phantom measurements. A monoexponential function of the form $f(x) \propto \exp(-\alpha|x|)$ is often chosen in PET. Although this approach does not take spatially inhomogeneous scattering into account, it has been shown that it leads to satisfying results in removing scattered events in 3D PET of phantom and human brain studies [90]. More accurate scatter models in both SPECT and PET take non-uniform scattering (i.e., spatially variant scatter functions and fractions) into account by using the acquired transmission data [59] [66]. Alternatively, the convolution-subtraction algorithm can also be applied to the PET data after reconstruction. This is done by independent reconstruction of both the measured data g_0 and the determined scatter distribution g_s and then finally subtracting the scatter image from the PET image [54].

Another way of using simple analytical functions for scatter correction of PET data is realized in tail-fitting methods. The basic idea here is the fact that measured coincidence events on lines that do not pass the scanned object must be scattered events. (Note that these methods do not work in SPECT, as scattered events appear to be confined to the scanned object in SPECT.) The measured data outside the object, the tail, is fitted with an analytical function such as a Gaussian which is then interpolated inside the object to get an estimate of the overall scatter distribution (Figure 5.21). This interpolation is legitimate due to the assumption that the scatter distribution is characterized as a low frequency function that does not strongly depend on high frequency distributions in the radiotracer distribution. The determined scatter distribution is then subtracted from the measured data. This method has been successfully implemented in human brain PET [22]. One of its advantages is that it takes scatter from outside the field of view into account which is a known problem especially in 3D PET. However, this method may fail in cases with asymmetric scatter distributions caused by non-uniform objects.

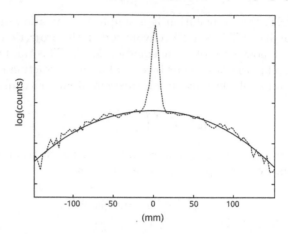

FIGURE 5.22: Fitting of 3D PET projection data (dashed line) with a Gaussian (solid line) to determine the scatter distribution in a scan of a line source inside a uniform cylinder.

5.5.3   Direct calculation methods

These methods are characterized by the usage of both emission and transmission data as well as a scanner model to calculate the scatter distribution according to basic Compton scatter physics (Klein–Nishina formula for the electron-photon cross section [47]). These methods are numerically more complex than the approaches described before; however, due to increasing computational speed available, they have become more and more important during the last years.

Most of these methods aim to determine the single scatter distribution (i.e., events where only one photon is scattered) as this is assumed to be the most prominent part of the overall scatter distribution (single scatter simulation, SSS) [72] [85]. Either appropriate scaling or integral transformations of this single scatter distribution is then taken as a good approximation of multiple scatter events. More recent algorithms even allow highly accurate direct determination of multiple scatter distributions [62]. Scatter from outside the field of view can be a problem in these methods; however, algorithms that take this explicitly into account have been developed and successfully implemented [1].

Monte Carlo simulation techniques have also been used to calculate the scatter distribution in emission tomography scans. In these approaches, an estimate of the actual tracer distribution is used in addition to a model of both the object and the scanner geometry to follow the fate of photons originating from voxels inside the object according to the tracer distribution [55]. A disadvantage of this approach is that scatter from outside the field of view is not taken into account. Additionally, the actual tracer distribution necessary for accurate scatter determination is not known, usually leading to a bias in scatter estimation. A refined approach taking this into account has been proposed by pre-correcting the emission data for scatter using the convolution-subtraction method [90].

5.5.4 Iterative reconstruction methods

Fast iterative image reconstruction algorithms, especially the ordered subsets expectation maximization (OSEM) algorithm, have become important competitors of analytical methods like filtered backprojection (FBP) during the last years. One of their properties is the different convergence rate of image features that have different spatial frequencies; low-frequency image components will converge substantially faster than high-frequency components as demonstrated by Fourier analysis of these algorithms [76]. As the scatter distribution in emission tomography imaging is assumed to have only low-frequency components, an early iteration in the reconstruction process may be a good estimate of the spatial distribution of scattered events. This approach has been successfully introduced in both SPECT [56] and PET [90] using one OSEM iteration.

5.6 Concluding remarks

Both PET and SPECT potentially offer the ability to absolutely quantify tracer distributions *in vivo*. For obvious reasons, this can be achieved only by a fundamental understanding and modeling of all relevant physical processes from radiation formation to radiation detection. Especially the progress in the field of new hybrid tomography systems, such as PET/CT and PET/MR scanners, necessitates the development of new correction methods (specifically methods correcting attenuation and scatter). Additionally, as the intrinsic spatial resolution of emission tomography systems is continuing to increase, proper correction for motion-related effects will be of growing importance in both clinical practice and pre-clinical research. Furthermore, ongoing increase of computational power allows more realistic models to be incorporated into the reconstruction process. Therefore, corrections for physical factors will remain an important part of future research in emission tomography.

References

[1] R. Accorsi, L.E. Adam, M.E. Werner, and J.S. Karp. Optimization of a fully 3D single scatter simulation algorithm for 3D PET. *Physics in Medicine and Biology*, 49:2577–2598, 2004.

[2] A. Ahmadian, M.R. Ay, J.H. Bidgoli, S. Sarkar, and H. Zaidi. Correction of oral contrast artifacts in CT-based attenuation correction of PET images using an automated segmentation algorithm. *European Journal of Nuclear Medicine and Molecular Imaging*, 35:1812–1823, 2008.

[3] A.M. Alessio, S. Kohlmyer, K. Branch, G. Chen, J. Caldwell, and P. Kinahan. Cine CT for attenuation correction in cardiac PET/CT. *Journal of Nuclear Medicine*, 48:794–801, 2007.

[4] G. Antoch, L.S. Freudenberg, T. Egelhof, J. Stattaus, W. Jentzen, J.F. Debatin, and A. Bockisch. Focal tracer uptake: a potential artifact in contrast-enhanced dual-modality PET/CT scans. *Journal of Nuclear Medicine*, 43:1339–1342, 2002.

[5] G. Antoch, H. Kuehl, J. Kanja, T.C. Lauenstein, H. Schneemann, E. Hauth, W. Jentzen, T. Beyer, S.C. Goehde, and J.F. Debatin. Dual-modality PET/CT scanning with negative oral contrast agent to avoid artifacts: introduction and evaluation. *Radiology*, 230:879–885, 2004.

[6] D.L. Bailey and S.R. Meikle. A convolution-subtraction scatter correction method for 3D PET. *Physics in Medicine and Biology*, 39:411–424, 1994.

[7] M. Bentourkia, P. Msaki, J. Cadorette, and R. Lecomte. Assessment of scatter components in multispectral PET imaging. *IEEE Nuclear Science Symposium Conference Record*, 3:1779–1783, 1993.

[8] A.K. Berthelsen, S. Holm, A. Loft, T.L. Klausen, F. Andersen, and L. Højgaard. PET/CT with intravenous contrast can be used for PET attenuation correction in cancer patients. *European Journal of Nuclear Medicine and Molecular Imaging*, 32:1167–1175, 2005.

[9] T. Beyer, G. Antoch, T. Blodgett, L.F. Freudenberg, T. Akhurst, and S. Mueller. Dual-modality PET/CT imaging: the effect of respiratory motion on combined image quality in clinical oncology. *European Journal of Nuclear Medicine and Molecular Imaging*, 30:588–596, 2003.

[10] T. Beyer, D.W. Townsend, T. Brun, P.E. Kinahan, M. Charron, R. Roddy, J. Jerin, J. Young, L. Byars, and R. Nutt. A combined PET/CT scanner for clinical oncology. *Journal of Nuclear Medicine*, 41:1369–1379, 2000.

[11] K. Bilger, L.E. Adam, and J.S. Karp. Segmented attenuation correction using Cs-137 single photon transmission. In *IEEE Nuclear Science Symposion and Medical Imaging Conference Record*, 2095–2099, 2001.

[12] S.C. Blankespoor, X. Xu, K. Kaiki, J.K. Brown, H.R. Tang, C.E. Cann, and B.H. Hasegawa. Attenuation correction of SPECT using X-ray CT on an emission-transmission CT system: myocardial perfusion assessment. *IEEE Transactions on Nuclear Science*, 43:2263–2274, 1996.

[13] L. Boucher, S. Rodrigue, R. Lecomte, and F. Bénard. Respiratory gating for 3-dimensional PET of the thorax: feasibility and initial results. *Journal of Nuclear Medicine*, 45:214–219, 2004.

[14] D. Brasse, P.E. Kinahan, C. Lartizien, C. Comtat, M. Casey, and C. Michel. Correction methods for random coincidences in fully 3D whole-body PET: Impact on data and image quality. *Journal of Nuclear Medicine*, 46:859–867, 2005.

[15] C. Burger, G. Goerres, S. Schoenes, A. Buck, A.H. Lonn, and G.K. von Schulthess. PET attenuation coefficients from CT images: experimental evaluation of the transformation of CT into PET 511-keV attenuation coefficients. *Eur Journal of Nuclear Medicine Mol Imaging*, 29:922–927, 2002.

[16] F. Büther, M. Dawood, L. Stegger, F. Wübbeling, M. Schäfers, O. Schober, and K.P. Schäfers. List mode-driven cardiac and respiratory gating in PET. *Journal of Nuclear Medicine*, 50:674–681, 2009.

[17] F. Büther, L. Stegger, M. Dawood, F. Range, M. Schäfers, R. Fischbach, T. Wichter, O. Schober, and K.P. Schäfers. Effective methods to correct contrast agent-induced errors in PET quantification in cardiac PET/CT. *Journal of Nuclear Medicine*, 48:1060–1068, 2007.

[18] J. Carney, T. Beyer, D. Brasse, J.T. Yap, and D.W. Townsend. CT-based attenuation correction for PET/CT scanners in the presence of contrast agent. *IEEE Nuclear Science Symposium Conference Record*, 3:1443–1446, 2002.

[19] R.E. Carson, M.E. Daube-Witherspoon, and M.V. Green. A method for postinjection PET transmission measurements with a rotating source. *Journal of Nuclear Medicine*, 29:1558–1567, 1988.

[20] M.E. Casey and E.J. Hoffman. Quantitation in positron emission computed tomography. 7. A technique to reduce noise in accidental coincidence measurements and coincidence efficiency calibration. *Journal of Computer Assisted Tomography*, 10:845–850, 1986.

[21] L.T. Chang. A method for attenuation correction in radionuclide computed tomography. *IEEE Transactions on Nuclear Science*, 25:638–643, 1978.

[22] S.R. Cherry and S.C. Huang. Effefcts of scatter on model parameter estimatesn 3D PET studies of the human brain. *IEEE Transactions on Nuclear Science*, 42:1147–1179, 1995.

[23] C. Cohade, M. Osman, Y. Nakamoto, L.T. Marshall, J.M. Links, E.K. Fishman, and R.L. Wahl. Initial experience with oral contrast in PET/CT: phantom and clinical studies. *Journal of Nuclear Medicine*, 44:412–416, 2003.

[24] M. Dawood, F. Büther, X. Jiang, and K.P. Schäfers. Respiratory motion correction in 3-D PET data with advanced optical flow algorithms. *IEEE Transactions on Medical Imaging*, 27:1164–1175, 2008.

[25] M. Dawood, F. Büther, N. Lang, O. Schober, and K.P. Schäfers. Respiratory gating in positron emission tomography: a comparison of different gating schemes. *Medical Physics*, 34:3067–3076, 2007.

[26] R.A. deKemp and C. Nahmias. Attenuation correction in PET using single photon transmission measurement. *Medical Physics*, 40:771–778, 1994.

[27] F.D. DiFilippo and R.C. Brunken. Do implanted pacemaker leads and ICD leads cause metal-related artifact in cardiac PET/CT? *Journal of Nuclear Medicine*, 46:436–443, 2005.

[28] O.C. Doerfler, A.J. Ruppert-Kohlmayr, P. Reittner, T. Hinterleitner, W. Petritsch, and D.H. Szolar. Helical CT of the small bowel with an alternative oral contrast material in patients with Crohn's disease. *Abdominal Imaging*, 28:313–318, 2003.

[29] C.E. Floyd, R.T. Jaszczak, K.L. Greer, and R.E. Coleman. Deconvolution of compton scatter in SPECT. *Journal of Nuclear Medicine*, 26:403–408, 1985.

[30] D. Gagnon, A. Todd-Pokropek, A. Arsenault, and G. Dupras. Introduction to holospectral imaging in nuclear medicine for scatter correction. *IEEE Trans Med Imag*, 8:245–250, 1989.

[31] G. Germano and E.J. Hoffman. Investigation of count rate and deadtime characteristics of a high resolution PET system. *J Comput Assist Tomogr*, 12:836–846, 1988.

[32] G.W. Goerres, E. Kamel, T.H. Heidelberg, M.T. Schwitter, C. Burger, and G.K. Schulthess. PET-CT image co-registration in the thorax: influence of respiration. *European Journal of Nuclear Medicine and Molecular Imaging*, 29:351–360, 2002.

[33] G.W. Goerres, S.I. Ziegler, C. Burger, T. Berthold, G.K. von Schulthess, and A. Buck. Artifacts at PET and PET/CT caused by metallic hip prosthetic material. *Radiology*, 226:577–584, 2003.

[34] S. Goetze, T.L. Brown, W.C. Lavely, Z. Zhang, and F.M. Bengel. Attenuation correction in myocardial perfusion SPECT/CT: effects of misregistration and value of reregistration. *Journal of Nuclear Medicine*, 48:1090–1095, 2007.

[35] S. Goetzel, T.L. Brown, W.C. Lavely, Z. Zhang, and F.M. Bengel. Attenuation correction in myocardial perfusion SPECT/CT: effects of misregistration and value of reregistration. *Journal of Nuclear Medicine*, 48:1090–1095, 2007.

[36] K.L. Gould, T. Pan, C. Loghin, N.P. Johnson, A. Guha, and S. Sdringola. Frequent diagnostic errors in cardiac PET/CT due to misregistration of CT attenuation and emission PET images: a definitive analysis of causes, consequences, and corrections. *Journal of Nuclear Medicine*, 48:1112–1121, 2007.

[37] S. Grootoonk, T.J. Spinks, D. Sashin, N.M. Spyrou, and T. Jones. Correction for scatter in 3D brain PET using a dual energy window method. *Physics in Medicine and Biology*, 41:2757–2774, 1996.

[38] J.H. Hubbell and S.M. Seltzer. Tables of x-ray mass attenuation coefficients and mass energy-absorption coefficients. http://physics.nist.gov/physrefdata/xraymasscoef/cover.html, 1996.

[39] R.H. Huesman, S.E. Derenzo, J.L. Cahoon, A.B. Geyer, W.W. Moses, D.C. Uber, T. Vuletich, and T.F. Budinger. Orbiting transmission source for positron emission tomography. *IEEE Transactions on Nuclear Science*, 35:735–739, 1988.

[40] R.J. Jaszczak, K.L. Greer, C.E. Floyd, C.G. Harris, and R.E. Coleman. Improved SPECT quantitation using compensation for scattered photons. *Journal of Nuclear Medicine*, 25:893–900, 1984.

[41] E.M. Kamel, C. Burger, A. Buck, G.K. von Schulthess, and G.W. Goerres. Impact of metallic dental implants on CT-based attenuation correction in a combined PET/CT scanner. *European Radiology*, 13:724–728, 2002.

[42] J.S. Karp, G. Muehllehner, H. Qu, and X.H. Yan. Singles transmission in volume imaging PET with a Cs-137 source. *Physics in Medicine and Biology*, 40:929–944, 1995.

[43] P.E. Kinahan, D.W. Townsend, T. Beyer, and D Sashin. Attenuation correction for a combined 3D PET/CT scanner. *Medical Physics*, 25:2046–2053, 1998.

[44] M.A. King, S.J. Glick, P.H. Pretorius, R.G. Wells, H.C. Gifford, M.V. Narayanan, and T. Farncombe. Attenuation, scatter, and spatial resolution compensation in SPECT. In *Emission Tomography. The fundamentals of PET and SPECT*. Elsevier Academic Press, 2004.

[45] M.A. King, G.J. Hademenos, and S.J. Glick. A dual-photopeak window method for scatter correction. *Journal of Nuclear Medicine*, 33:605–612, 1992.

[46] G.J. Klein, B.W. Reutter, M.W. Ho, J.H. Reed, and R.H. Huesman. Real-time system for respiratory-cardiac gating in positron tomography. *IEEE Transactions on Nuclear Science*, 45:2139–2143, 1998.

[47] O. Klein and Y. Nishina. Über die Streuung von Strahlung durch freie Elektronen nach der neuen relativistischen Quantenmechanik nach Dirac. *Z Phys*, 52:853–868, 1929.

[48] G.F. Knoll. *Radiation Detection and Measurement*. John Wiley & Sons, New York, NY, USA, 1999.

[49] P. Koepfli, T.F. Hany, C.A. Wyss, M. Namdar, C. Burger, A.V. Konstantinidis, T. Berthold, G.K. von Schulthess, and P.A. Kaufmann. CT attenuation correction for myocardial perfusion quantification using a PET/CT hybrid scanner. *Journal of Nuclear Medicine*, 45:537–542, 2004.

[50] F. Lamare, T. Cresson, J. Savean, C. Cheze-Le Rest, A.J. Reader, and D. Visvikis. Respiratory motion correction for PET oncology applications using affine transformation of list mode data. *Physics in Medicine and Biology*, 52:121–140, 2007.

[51] N. Lang, M. Dawood, F. Büther, O. Schober, M. Schäfers, and K. Schäfers. Organ movement reduction in PET/CT using dual-gated list mode acquisition. *Z Med Phys*, 16:93–100, 2006.

[52] R. Lautamäki, T.L.Y. Brown, J. Merrill, and F.M. Bengel. CT-based attenuation correction in 82Rb-myocardial perfusion PET-CT: incidence of misalignment and effect on regional tracer distribution. *European Journal of Nuclear Medicine and Molecular Imaging*, 35:305–310, 2008.

[53] C. Lemmens, D. Faul, J. Hamill, S. Stroobants, and J. Nuyts. Suppression of metal streak artifacts in CT using MAP reconstruction procedure. *IEEE Nuclear Science Symposium Conference Record*, 6:3431–3437, 2006.

[54] M.J. Lercher and K. Wienhard. Scatter correction in 3D PET. *IEEE Transactions on Medical Imaging*, 13:649–657, 1994.

[55] C.S. Levin, M. Dahlbom, and E.J. Hoffman. A Monte Carlo correction for the effect of compton scattering in 73-d pet. *IEEE Transactions on Nuclear Science*, 42(4):1185, 1995.

[56] Z. Liu, T. Obi, M. Yamaguchi, and N. Ohyama. Fast estimation of scatter components using the ordered subsets expectation maximization algorithm for scatter compensation. *Optical Review*, 6:415–423, 1999.

[57] L. Livieratos, K. Rajappan, L. Stegger, K. Schäfers, D.L. Bailey, and P.G. Camici. Respiratory gating of cardiac PET data in list-mode acquisition. *European Journal of Nuclear Medicine and Molecular Imaging*, 33:584–588, 2006.

[58] L. Livieratos, L. Stegger, P.M. Bloomfield, K. Schäfers, D.L. Bailey, and P.G. Camici. Rigid-body transformation of list mode projection data for respiratory motion correction in cardiac PET. *Physics in Medicine and Biology*, 50:3313–3322, 2005.

[59] M. Ljungberg and S.E. Strand. Scatter and attenuation correction in SPECT using density maps and monte carlo simulated scatter functions. *Journal of Nuclear Medicine*, 31:1560–1567, 1990.

[60] A. Lupi, M. Zaroccolo, M. Salgarello, V. Malfatti, and P. Zanco. The effect of 18F-FDG-PET/CT respiratory gating on detected metabolic activity in lung lesions. *Annals of Nuclear Medicine*, 23:191–196, 2009.

[61] A.H. Mahnken, R. Raupach, J.E. Wildberger, B. Jung, Heussen N., T.G. Flohr, R.W. Günther, and S. Schaller. A new algorithm for metal artifact reduction in computed tomography: in vitro and in vivo evaluation after total hip replacement. *Investigative Radiology*, 12:769–775, 2003.

[62] P.J. Markiewicz, M. Tamal, P.J. Julyan, D.L. Hastings, and A.J. Reader. High accuracy multiple scatter modelling for 3D whole body PET. *Physics in Medicine and Biology*, 52:829–847, 2007.

[63] A. Martinez-Möller, M. Souvatzoglou, N. Navab, M. Schwaiger, and S.G. Nekolla. Artifacts from misaligned CT in cardiac perfusion PET/CT studies: frequency, effects, and potential solutions. *Journal of Nuclear Medicine*, 48:188–193, 2007.

[64] O. Mawlawi, J.J. Erasmus, R.F. Munden, T. Pan, A.E. Knight, H.A. Macapinlac, D.A. Podoloff, and M. Chasen. Quantifying the effect of IV contrast media on integrated PET/CT: clinical evaluation. *American Journal of Roentgenology*, 186:308–319, 2006.

[65] S.R. Meikle, M. Dahlbohm, and S.R. Cherry. Attenuation correction using count-limited transmission data in positron emission tomography. *Journal of Nuclear Medicine*, 34:143–150, 1993.

[66] S.R. Meikle, B.F. Hutton, and D.L. Bailey. A transmission dependent method for scatter correction in SPECT. *Journal of Nuclear Medicine*, 35:360–367, 1994.

[67] C. Nahmias, D. Lemmens, C.and Faul, E. Carlson, M. Long, T. Blodgett, J. Nuyts, and D. Townsend. Does reducing CT artifacts from dental implants influence the PET interpretation in PET/CT studies of oral cancer and head and neck cancer? *Journal of Nuclear Medicine*, 49:1047–1052, 2008.

[68] S.A. Nehmeh, Y.E. Erdi, H. Kalaigian, K.S. Kolbert, T. Pan, H. Yeung, O. Squire, A. Sinha, S.M. Larson, and J.L. Humm. Correction for oral contrast artifacts in CT attenuation-corrected PET images obtained by combined PET/CT. *Journal of Nuclear Medicine*, 44:1940–1944, 2003.

[69] S.A. Nehmeh, Y.E. Erdi, C.C. Ling, K.E. Rosenzweig, H. Schoder, S.M. Larson, H.A. Macapinlac, O.D. Squire, and J.L. Humm. Effect of respiratory gating on quantifying PET images of lung cancer. *Journal of Nuclear Medicine*, 43:876–881, 2002.

[70] S.A. Nehmeh, Y.E. Erdi, T. Pan, A. Pevsner, K.E. Rosenzweig, E. Yorke, G.S. Mageras, H. Schoder, P. Vernon, O. Squire, H. Mostafavi, S.M. Larson, and J.L. Humm. Four-dimensional (4D) PET/CT imaging of the thorax. *Medical Physics*, 31:3179–3186, 2004.

[71] J.A. Nye, F. Esteves, and J.R. Votaw. Minimizing artifacts resulting from respiratory and cardiac motion by optimization of the transmission scan in cardiac PET/CT. *Medical Physics*, 34:1901–1906, 2007.

[72] J.M. Ollinger. Model-based scatter correction for fully 3D PET. *Physics in Medicine and Biology*, 41:153–176, 1996.

[73] M.M. Osman, C. Cohade, Y. Nakamoto, and R.L. Wahl. Respiratory motion artifacts on PET emission images obtained using CT attenuation correction on PET-CT. *European Journal of Nuclear Medicine and Molecular Imaging*, 30:603–605, 2003.

[74] M.R. Palmer, J.G. Rogers, M Bergstrom, M.P. Bedooes, and B.D. Pate. Transmission profile filtering for positron emission tomography. *IEEE Transactions on Nuclear Science*, 33:478–481, 1986.

[75] T. Pan, O. Mawlawi, S.A. Nehmeh, Y.F. Erdi, D. Luo, H.H. Liu, R. Castillo, R. Mohan, Z. Liao, and H.A. Macapinlac. Attenuation correction of PET images with respiration-averaged CT images in PET/CT. *Journal of Nuclear Medicine*, 46:1481–1487, 2005.

[76] T.S. Pan and Yagle A.E. Numerical study of multigrid implementations of some iterative reconstruction algorithms. *IEEE Transactions on Medical Imaging*, 10:572–588, 1991.

[77] L.M. Popescu, R.M. Lewitt, S. Matej, and J.S. Karp. PET energy-based scatter estimation and image reconstruction with energy-dependent corrections. *Physics in Medicine and Biology*, 51:2919–2937, 2006.

[78] N.S. Rehfeld, B.J. Heismann, J. Kupferschläger, P. Aschoff, G. Christ, A.C. Pfannenberg, and B.J. Pichler. Single and dual energy attenuation correction in PET/CT in the presence of iodine based contrast agents. *Medical Physics*, 35:1959–1969, 2008.

[79] K.P. Schäfers, R. Raupach, and T. Beyer. Combined 18F-FDG-PET/CT imaging of the head and neck—an approach to metal artifact correction. *Nuklearmedizin*, 45:219–222, 2006.

[80] P.J. Schleyer, M.J. O'Doherty, S.F. Barrington, and P.K. Marsden. Retrospective data-driven respiratory gating for PET/CT. *Physics in Medicine and Biology*, 54:1935–1950, 2009.

[81] L. Shao, R. Freifelder, and J.S. Karp. Triple energy window scatter correction technique in PET. *IEEE Transactions on Medical Imaging*, 4:641–648, 1994.

[82] M. Souvatzoglou, F. Bengel, R. Busch, C. Kruschke, H. Fernolendt, Lee D., M. Schwaiger, and S.G. Nekolla. Attenuation correction in cardiac PET/CT with three different CT protocols: a comparison with conventional PET. *European Journal of Nuclear Medicine and Molecular Imaging*, 34:1991–2000, 2007.

[83] H.R. Tang, J.K. Brown, A.J. da Silva, K.K. Matthay, D.C. Price, J.P. Huberty, R.A. Hawkins, and B.H. Hasegawa. Implementation of a combined X-ray CT-scintillation camera imaging system for localizing and measuring radionuclide uptake: experiments in phantoms and patients. *IEEE Transactions on Nuclear Science*, 46:551–557, 1999.

[84] D.W. Townsend, J.P.J. Carney, J.T. Yap, and N.C Hall. PET/CT today and tomorrow. *Journal of Nuclear Medicine*, 45:4S–14S, 2005.

[85] C.C. Watson. New, faster, image-based scatter correction for 3D PET. *IEEE Transactions on Nuclear Science*, 47:1587–1594, 2000.

[86] C.C. Watson, D.W. Townsend, and B. Bendriem. PET/CT systems. In *Emission Tomography. The fundamentals of PET and SPECT*. Elsevier Academic Press, 2004.

[87] S. Weng and B. Bettinadi. An automatic segmentation method for fast imaging in PET. *Nuclear Science and Techniques*, 4:114–119, 1993.

[88] M. Xu, W.K. Luk, P.D. Cutler, and W.N. Digby. Local threshold for segmented attenuation correction of PET imaging of the thorax. *IEEE Transactions on Nuclear Science*, 41:1532–1537, 1994.

[89] Y. Yau, W. Chan, Y. Tam, P. Vernon, S. Wong, M. Coel, and S.K. Chu. Application of intravenous contrast in PET/CT: Does it really introduce significant attenuation correction error? *Journal of Nuclear Medicine*, 46:283–291, 2005.

[90] H. Zaidi. Comparative evaluation of scatter correction techniques in 3D positron emission tomography. *European Journal of Nuclear Medicine*, 27:1813–1826, 2000.

[91] H. Zaidi. Scatter modelling and correction strategies in fully 3-D PET. *Nuclear Medicine Communications*, 22:1181–1184, 2001.

[92] H. Zaidi and B. Hasegawa. Determination of the attenuation map in emission tomography. *Journal of Nuclear Medicine*, 44:291–315, 2002.

Chapter 6

Corrections for Scanner-Related Factors

Marc Huismann
Department of Nuclear Medicine & PET Research, VU University Medical Center, Amsterdam, the Netherlands

6.1   Positron emission tomography 105
    6.1.1   Introduction 105
    6.1.2   Data normalization 107
    6.1.3   Noise equivalent count rates 108
    6.1.4   System dead time 108
    6.1.5   Partial volume 110
6.2   Single photon emission computed tomography 112
    6.2.1   Linearity, center of rotation, and whole body imaging .. 112
    6.2.2   Motion correction 114
References 115

6.1 Positron emission tomography

6.1.1 Introduction

In PET a positron annihilation event results in two 511 keV photons. Due to conservation of momentum these photons are emitted in approximately 180-degree opposing directions. When detected as a prompt coincidence, the detectors in which the photons were detected define a so-called line of response (LOR; actually it is a tube of response, the volume of which is determined by the distance between the detectors and the area of the detector surface).

The term detector is used to refer to a scintillation crystal that can be individually identified. Usually, Anger-logic readout of block detectors leads to the capability to address small individual crystals (e.g., 4 x 4 x 22 mm^3 LYSO crystals in the Philips Gemini TF PET-CT scanner [18], 4 x 4 x 20 mm^3 LSO crystals in the Siemens Biograph 16 HI-REZ PET-CT scanner [3] and 4.7 x 6.3 x 30 mm^3 BGO crystals in the GE Discovery STE PET-CT scanner [12]). However, even a continuous distribution of scintillation crystals over a ring of detectors leads to an inhomogeneous density distribution of LOR in

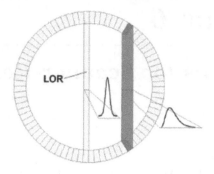

FIGURE 6.1: A schematic representation of the depth-of-interaction effect in 2D-PET. Taken from [14].

the field of view (FOV) of a PET scanner. This is caused by the depth-of-interaction effect (see Figure 6.1 for a schematic representation of this effect in 2D, taken from [14]) or through the presence of gaps in the FOV (in the case when the detectors are distributed over panels). In the axial direction, LOR density will be less at the edges than in the middle of the FOV.

Traditionally, raw data consisted of the total number of measured prompts and randoms acquired during acquisition per LOR. This data was rebinned into a sinogram. The rebinning step is necessary in order to produce a consistent and complete set of projection data that is put into the reconstruction code. In order to put the measured data into a sinogram a trade-off between counting statistics and sampling detail needs to be found (see, e.g., [7]). In general the choices have been made by the manufacturer of the scanner. User interaction is usually limited or even completely absent. The following quantities can be discussed:

- Not only coincidences within the same ring are taken into account, but coincidences between adjacent rings as well. For an N ring scanner a total of N direct $+$ $(N$-$1)$ indirect planes can thus be defined, with an axial sampling equal to half the detector spacing. The term span is used to indicate the amount of axial data combinations; the number refers to the sum of the number of planes contributing to a direct plane and the number of (cross-) planes contributing to an indirect ('interpolated') frame.

- Alternatively, radial bins can be combined in order to improve sinogram sampling. A mashing factor of 1 implies that two angular samples are taken together and attributed to the average angle of the two.

- The maximum ring difference indicates the minimum angle between the z-axis and a line of response that is still being taken into the sinogram (an in-plane line of response lies perpendicularly to the z-axis). Usually, axially angled LOR are taken together into segments (segment 0 comprising the non-axially angled LOR). Within a segment various axial angles are combined. For an increasing axial angle a decreasing z-coverage of the FOV results.

The newest generation PET scanners acquire raw data in list-mode format. It is possible to convert list-mode data into sinograms, but newer approaches are explored that might not rely on the stringent requirements of a complete set of projection data that accounted for some of the trade-offs mentioned above [15]. However, these techniques did not enter clinical practice yet and will not be discussed.

After a prompt sinogram has been obtained, corrections need to be made in order to allow for a reconstruction of the measured data that yield quantitatively accurate data. Most corrections are carried out in the form of multiplications or summations/subtractions of estimated or measured sinograms containing correction factors (i.e., randoms, attenuation, and scatter as discussed in chapter 5).

6.1.2 Data normalization

In image reconstruction it is assumed that every photon that is incident on a crystal surface has the same probability of being detected. In practice, this probability differs from crystal to crystal, for various reasons:

- the growth of single crystal ingots that are being cut subsequently into detector blocks is a difficult process that does not allow for a complete absence of differences in crystal characteristics (light output, decay time, etc.).

- performance tolerances on the signal processing chain (e.g., photomultipliers, ADCs, and possible drift in these components) will lead to small changes in the energy spectrum (number of channels in the energy window of interest) and (stability of the) calibration of the energy scale.

- the amount of light detected for a photon interacting in the middle of a block detector is higher than for a photon interacting at the edge of a block detector (not to speak of scatter within the detector).

- the angle of incidence between the incoming photon and the crystal surface normal. For larger angles a smaller path length and, subsequently, a smaller interaction probability exists.

- a sinogram bin relating to a direct plane will have a lower sensitivity than a sinogram bin relating to an indirect plane (roughly twice as low for a span 3 sinogram).

- timing calibration and its stability determine whether a photon pair is considered to be due to a positron emission or due to a random decay process.

Some of these factors are due to the limitations of current instrumentation (like crystal quality, timing and energy calibration). The influence of these factors cannot be predicted. Other factors are related to the detector geometry. These can be predicted on the basis of calculations of simulations. In either case, a correction for these effects needs to be carried out. This is the normalization correction. Generally speaking, two approaches can be taken. In direct normalization all lines of response do see an activity distribution, and data are acquired (at low count rates to minimize the effects of randoms and scatters) for a sufficiently long time to allow for some statistical certainty on the normalization factors, which will be equal to the reciprocal of the number of acquired counts. Assuming Poisson statistics, an error of 10% would require at least 100 counts in every sinogram bin (normalization factors are placed in a sinogram and multiplied with the measured trues sinogram). For a modern scanner with a large number of crystals this can imply a prohibitive amount of time needed for normalization. Therefore, component-based normalization schemes have become popular in which as many components of the normalization as possible are precomputed (i.e., geometrical factors). The crystal efficiencies are estimated from a scan on, e.g., a cylindrical phantom, where a particular crystal is acquiring data in coincide with a sum of opposite detectors. This increases count statistics, and the assumption is that the total number of coincidences measured is still a good measure for the crystal efficiency.

6.1.3 Noise equivalent count rates

The concept of noise equivalent count rates addresses the fact that at increased count rates the number of randoms and scatter events increases as well. In the NEMA recommendations for the performance evaluation of PET scanners the determination of NEC curves is therefore prescribed. In addition, the dead time characteristics of the system need to be taken into account.

6.1.4 System dead time

The term *dead time* refers to the finite time a general pulse processing system needs in order to process an event [10, 11]. During this processing time the system cannot process a following incoming event. There are two ways in which the system can cope with this second (or higher order) event (see Figure 6.2):

- the system neglects the event, and after processing the first event the next-coming event will be processed. Such a system is called non-paralyzable. At increasing count rates the system will at first linearly increase the number of processed events. If the count rate becomes of

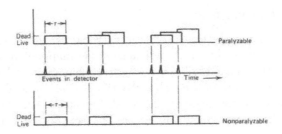

FIGURE 6.2: During dead time the system cannot process a following incoming event. Taken from [10].

the order of (1/processing time) the chance of pulse pile up (two or more events arriving during the processing time) increases. For a nonparalyzable system the count rate response will start to deviate from linearity, and it will level off to a maximum count rate.

- the system processes piled-up events, increasing the effective processing time each time a pile-up event is added. After onset of non-linear behavior of the detected count rate versus the expected count rate (based on a linear increase with radioactivity in the field of view) the count rate will level off, but may start decreasing as well. A maximum in the expected vs. measured count rate curve indicates the presence of a paralyzable pulse processing system.

Although the term is known and can be found in many pulse processing systems, it has been of major concern in 3D PET. Correction for dead time is usually provided by reference to a lookup table that links the measured singles rate to a dead-time correction factor. In practice a 10% dead-time correction is taken to constitute an upper limit to the count rate that can be processed in a study that will be quantified.

Recently, interesting developments in preclinical systems have been described: the Quicksilver electronics platform for the Siemens Inveon scanner family has significantly reduced dead-time–related problems, and the degree of multiplexing in various experimental systems has also been reflected in the reported count rate capability of the systems (e.g., MADPET, SiPM).

For example, cardiac studies with Rb are currently limited with respect to the injected dose by the dead-time characteristics of the PET scanner. On the other hand, increasing the level at which 10% dead-time correction is applied is useful only in case this leads to an improved NEC rate, which might not be the case for a situation in which considerable scatter is present in the system.

6.1.5　Partial volume

Partial volume corrections are necessary due to the discrete nature with which the true distribution of radioactivity is mapped into an image. Discretization of the imaged volume into voxels that may be large with respect to the spatial variation in the radioactivity distribution leads to an averaging of this variation on the scale of the voxel dimension. Another length scale is provided by the spatial resolution of the PET scanner. This macro parameter depends on a number of characteristics:

- geometrical arrangement of the detectors (e.g., the axial dimension of a crystal element sets a lower limit to finest axial spacing in sinogram space). This arrangement also introduces the typical spatial variation of spatial resolution over the FOV of the scanner.

- the physics of positron annihilation; due to annihilation photon non-collinearity, a larger scanner bore will lead to a decreased spatial resolution of the scanner.

- the possibility to determine depth of interaction within a crystal; by using a binary depth of interaction scheme a recorded event can be attributed to either the surface or the back part of a crystal element (typically 1 cm long). Depth of interaction can be measured by defining a physically separated front and back layer for each original crystal (by combining two different scintillators, or having a separate readout for two stacked crystals that both are half the length of an original crystal), double readout of a single crystal (front and back side) and so on. This effectively reduces the distance between lines of responses, enabling a finer sinogram sampling.

- choices for span, ring difference and radial and angular mashing factors; increasing those factors generally leads to increased spatial resolution of the scanner, but they may be necessitated by count statistical considerations.

- the reconstruction algorithm.

The NEMA performance evaluation protocol [2] describes a method to determine spatial resolution of a PET camera. The source should be smaller than 4 times the expected FWHM spatial resolution of the scanner. The activity should be low enough to make contributions of randoms and scatters to the measured data negligible. Reconstruction should be done via filtered back projection (a non-iterative reconstruction method) on a voxel grid that provides at least 5 pixels on the FWHM. Profiles through the maximum of the measured radioactivity distribution should be summed over two FWHM in the two directions perpendicular to the direction in which the FWHM is measured. The maximum activity should be determined from a parabolic fit to the measured data, and the FWHM and FWTM values of the spatial resolution should be determined from a linear interpolation between the neighboring measured data points. The alternative method of line of point transfer functions is hardly used anymore. Data should be obtained for various locations in the FOV of the PET scanner.

As a practical issue many preclinical scanners do not provide an FBP reconstruction algorithm. This has led to a highly varying body of literature on the actually obtainable spatial resolution for a given scanner by applying all kinds of iterative reconstruction algorithms. It is important to note that iteratively reconstructed point source measurements against a cold background do depend strongly on the number of iterations or other tweaking factors. At least a non-zero background concentration should be present in the measurement of a spatial resolution based on an iterative reconstruction technique.

The finite spatial resolution of PET scanners (clinically ~6 mm, preclinically ~2 mm) leads to another effect that needs to be considered if quantitative data analysis is to follow the reconstruction of PET images. Radioactivity in structures with dimensions smaller than 2 × FWHM spatial resolution of the scanner will be affected by the partial volume effect (PVE). In these structures the measured radioactivity will be spread out over a volume roughly equal to a cylinder with a diameter of 2 × FWHM in the in-plane direction and a length of 2 × FWHM in the axial direction. This leads to an underestimation of the local radioactivity within the cylinder. Furthermore, the background activity surrounding the structure will influence the measured radioactivity concentration (via spill in for a structure in a background with a higher radioactivity concentration, or via spill out for a structure in a background with a lower radioactivity concentration). As an example, quantification of glucose metabolism in the myocardium is hindered by the high concentration of tracer in the blood pool at the start of the measurement (this activity spills into the myocardium). At the end of the measurement, the activity in the myocardium spills into the blood pool, which at that point in time will have a low remaining activity concentration. Phantom experiments with spheres of various sizes allow for the calculation of recovery coefficients that allow for compensation of these effects in spherical objects. But in the case when the tumor shape is not known, they are of limited usefulness. In a recent paper, various strategies to correct for the PVE were evaluated [9]. Corrections can be applied at various stages in the data analysis process: during reconstruction (see Chapter 3), during kinetic modeling or on the reconstructed images themselves. PVC improves accuracy of SUV without decreasing (clinical) test-retest variability significantly and it has a small, but significant effect on observed tumor responses. Reconstruction-based PVC outperforms image-based methods, but requires dedicated reconstruction software. Image-based methods are good alternatives because of their ease of implementation and their similar performance in clinical studies. For more detailed information on the methods that were evaluated one could have a look at the references of this chapter as well as in [16]. Furthermore PVE correction approaches have developed in different directions for different applications. Since MRI data is usually available for brain imaging, this information has been taken into account (see, e.g., [19, 20]). For tumor imaging, a recent overview is provided in [17].

6.2 Single photon emission computed tomography

SPECT is based on the detection of photons emitted by radiotracers that were injected into a patient. The basic detection block is a gamma detector head, consisting of a scintillator crystal, a light guide and an array of photomultiplier tubes. In front of the scintillator crystal a collimator allows for detection of only those photons that travel in a direction (almost) perpendicular to the gamma camera surface.

6.2.1 Linearity, center of rotation, and whole body imaging

Compensation methods in SPECT for the correction of scanner-related factors include methods to ensure

1. uniformity and linearity of the response of a gamma camera head to incoming photons (see, e.g., [6]);

2. coinciding definitions of the center of the field of view with the center of rotation of a gamma camera head;

3. consistency between bed motion and electronic windowing of a camera head in whole body acquisition settings.

Most of these correction methods are built into the scanner by the vendor. Details are not always accessible, but procedures for quality assurance have been outlined, for example, in a NEMA standard [1].

Per point 1 above, Linearity correction for a camera head ensures that the response of the camera to an incoming photon is as independent as possible on the location of incidence of the photon on the camera head surface (x, y location). However, in practice image nonlinearity is present in the raw measured data. Straight-line objects appear as curved-line images, leading to pincushion and/or barrel distortion. Physical reasons for this observation in properly functioning hardware are differences in sensitivity among PM tubes and nonuniformities in optical light guides. Malfunctions of these hardware components can also lead to image non-linearity.

Exposing a detector crystal to a uniform flux of photons produces a flood-field image. Even a properly functioning gamma camera head will show variations in the image intensity, which may be of the order of $\pm 5\%$ or more. Intrinsic flood-field images are acquired without collimator, usually with a point source positioned at a distance of 4–5 times the gamma camera head diameter. Extrinsic flood-field images are acquired with the collimator in place using a disk or thin flood phantom that covers the area of the detector.

A number of non-uniformity patterns is typically observed. One of them is called edge-packing, and is related to geometric factors enhancing the PMT output for edge portions of the gamma camera head. For this reason a distinction is made between the total field of view (TFOV) and the useful field of view (UFOV), where the latter excludes the edge region of the gamma camera head (typically ∼5 cm). Since the PMT response to incoming radiation is the main source for image non-uniformities, the uniformity is energy dependent.

Geometric sources of image non-uniformities may be compensated for by the use of appropriate lookup tables, based on high statistic scans that are acquired 2–4 times per year by the user. In practice, image non-uniformity is dealt with in a regular quality control program. Acceptable values are <3% in the UFOV.

Per point 2 above, misalignment between the electronic and mechanical axes of rotation will result in artifact generation and image degradation during reconstruction of projections obtained from one or more rotating gamma camera heads. Image reconstruction in which the center of rotation (COR) error for a given slice is larger than 0.5 pixels will lead to image degradation and artifacts [5]. Since alignment of these axes is assumed during image reconstruction, possible (mechanically induced) misalignments need to be corrected for. Most SPECT cameras inhere a quality control measurement that allows for an evaluation of the presence and magnitude of a possible misalignment. If the electronic and mechanical axes of rotation are aligned, then a single COR measurement is applicable for the complete FOV, but for verification of the alignment multiple (4–5) point sources need to be measured under various angles. The consistency of the resulting linograms can be assessed. Corrections for small misalignments are implemented in image reconstruction by shifting certain projection profiles before incorporation of the data into the reconstruction process.

Per point 3 above, whole body gamma camera imaging can usually be done in one run, given that a typical gamma camera head measures 50 x 40 cm. Whole body imaging makes uses of an electronic opening and closing of the gamma camera head FOV as well as physical movement of the table. The various actions need to be timed correctly, and the velocities need to be tuned such that for a constant radioactivity concentration on the table a constant radioactivity concentration is visible in the reconstructed whole body image. By placing a flood source on a gamma camera head the radioactivity concentration profile obtained from a whole body acquisition should be constant.

A recent trend in SPECT hardware incorporates a CT scanner, which allows for delineation of structures, improves localization of uptake and provides information needed for attenuation correction [13, 4]. Since both components of such a scanner contain established technology, the major factor that needs to be taken into account in the context of scanner-related correction factors is that for the translation between the CFOV of the SPECT and the CT scanner. Furthermore, differences in the diameter of the FOV between the two modalities may influence reconstruction accuracy. For myocardial perfusion SPECT/CT this issue is addressed in [8].

Furthermore, the intrinsic dependency of collimator resolution on source-to-collimator distance (see Figure 6.3) leads to the requirement that a patient is placed as close to the collimator as possible (within constraints of patient comfort, and sometimes the wish to keep this distance as equal as possible during a whole body acquisition).

6.2.2 Motion correction

For certain applications such as cardiac SPECT-CT imaging, advanced methods are available for compensation of motion. One form is compensation for cardiac motion by taking a measured ECG signal into account to sort measured data over various phases of the cardiac cycle (e.g., end-systolic or end-diastolic phases). Another form is an automated detection of patient motion during acquisition. If, for instance, during rotation of the gamma camera heads around the patient one of the projections is measured on a patient that moved between the former and the latter projection, this can be detected as well as automatically corrected (see Figure 6.4).

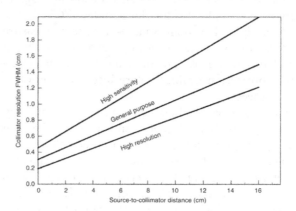

FIGURE 6.3: Collimator resolution versus source-to-collimator distance for three different collimators. Taken from [6].

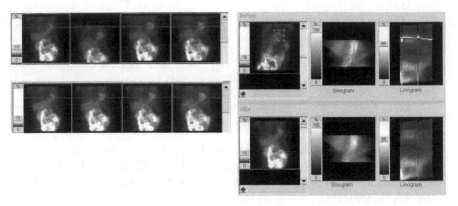

FIGURE 6.4: On the left, individual projections are shown with the top row before and the bottom row after automatic motion correction. Please note the change in position of the patient during the acquisition of the second projection shown. Correction is based on a center of mass calculation on a part of the projection, where discontinuities in the sinogram and linogram can be detected.

References

[1] NEMA Standards Publication NU 1-2007. Performance measurements of Gamma cameras. Technical report, National Electrical Manufacturers Association, Washington, DC, 2007.

[2] NEMA Standards Publication NU 2-2007. Performance measurements of positron emission tomographs. Technical report, National Electrical Manufacturers Association, Washington, DC, 2007.

[3] M. Brambilla, C. Secco, M. Dominietto, R. Matheoud, G. Sacchetti, and E. Inglese. Performance characteristics obtained for a new 3-dimensional lutetium oxyorthosilicate-based whole-body PET/CT scanner with the National Electrical Manufacturers Association NU 2-2001 standard. *Journal of Nuclear Medicine*, 46(12):2083, 2005.

[4] A.K. Buck, S. Nekolla, S. Ziegler, A. Beer, B.J. Krause, K. Herrmann, K. Scheidhauer, H.J. Wester, E.J. Rummeny, M. Schwaiger, et al. SPECT/CT. *Journal of Nuclear Medicine*, 49(8):1305, 2008.

[5] M.D. Cerqueira, D. Matsuoka, J.L. Ritchie, and G.D. Harp. The influence of collimators on SPECT center of rotation measurements: artifact gen-

eration and acceptance testing. *Journal of Nuclear Medicine*, 29(8):1393, 1988.

[6] S.R. Cherry, J.A. Sorenson, and M.E. Phelps. *Physics in Nuclear Medicine*. Saunders, 3rd edn edition, 2003.

[7] F.H. Fahey. Data acquisition in PET imaging. *Journal of Nuclear Medicine Technology*, 30(2):39, 2002.

[8] S. Goetze, T.L. Brown, W.C. Lavely, Z. Zhang, and F.M. Bengel. Attenuation correction in myocardial perfusion SPECT/CT: effects of misregistration and value of reregistration. *Journal of Nuclear Medicine*, 48(7):1090, 2007.

[9] N. J. Hoetjes, F. H. P. van Velden, O. S. Hoekstra, C. J. Hoekstra, N. C. Krak, A. A. Lammertsma, and R. Boellaard. Partial volume correction strategies for quantitative FDG PET in oncology. *Eur J Nucl Med Mol Imaging*, 37(9):1679–87, August 2010.

[10] G.F. Knoll. *Radiation Detection and Measurement* New York: John Wiley and Sons, 1979.

[11] W.R. Leo. *Techniques for Nuclear and Particle Physics Experiments*. Berlin: Springer-Verlag, 2nd edition, 2000.

[12] L.R. MacDonald, R.E. Schmitz, A.M. Alessio, S.D. Wollenweber, C.W. Stearns, A. Ganin, R.L. Harrison, T.K. Lewellen, and P.E. Kinahan. Measured count-rate performance of the Discovery STE PET/CT scanner in 2D, 3D and partial collimation acquisition modes. *Physics in Medicine and Biology*, 53:3723, 2008.

[13] J.A. Patton and T.G. Turkington. SPECT/CT physical principles and attenuation correction. *Journal of Nuclear Medicine Technology*, 36(1):1, 2008.

[14] B.J. Pichler, H.F. Wehrl, and M.S. Judenhofer. Latest advances in molecular imaging instrumentation. *Journal of Nuclear Medicine*, 49(Suppl_2):5S, 2008.

[15] A.J. Reader. The promise of new PET image reconstruction. *Physica Medica*, 24(2):49–56, 2008.

[16] O. Rousset, A. Rahmim, A. Alavi, and H. Zaidi. Partial volume correction strategies in PET. *PET Clinics*, 2(2):235–249, 2007.

[17] M. Soret, S.L. Bacharach, and I. Buvat. Partial-volume effect in PET tumor imaging. *Journal of Nuclear Medicine*, 48(6):932, 2007.

[18] S. Surti, A. Kuhn, M.E. Werner, A.E. Perkins, J. Kolthammer, and J.S. Karp. Performance of Philips Gemini TF PET/CT scanner with special consideration for its time-of-flight imaging capabilities. *Journal of Nuclear Medicine*, 48(3):471, 2007.

[19] C. Svarer, K. Madsen, S.G. Hasselbalch, L.H. Pinborg, S. Haugbøl, V.G. Frøkjær, S. Holm, O.B. Paulson, and G.M. Knudsen. MR-based automatic delineation of volumes of interest in human brain PET images using probability maps. *Neuroimage*, 24(4):969–979, 2005.

[20] H. Zaidi, M.L. Montandon, and S. Meikle. Strategies for attenuation compensation in neurological PET studies. *Neuroimage*, 34(2):518–541, 2007.

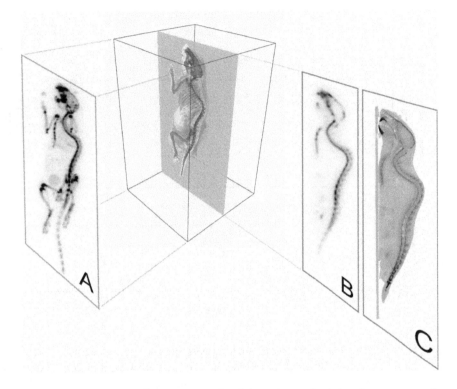

FIGURE 1.1: Two different biomedical imaging principles demonstrated on a ^{18}F-fluorid PET bone scan of a mouse: (A) projection image showing the maximum intensity projection; (B) single PET slice out of the 3D tomographic volume; (C) same slice but fused with CT data.

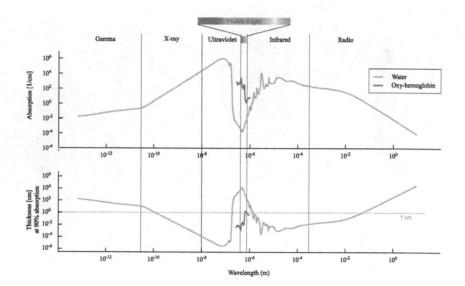

FIGURE 1.2: The electromagnetic spectrum showing the absorption coefficient in water at different wavelengths. Both, gamma radiation and visible light have low attenuation in water; oxy-hemoglobin adds a considerable attenuation (compiled from [1, 2, 3]).

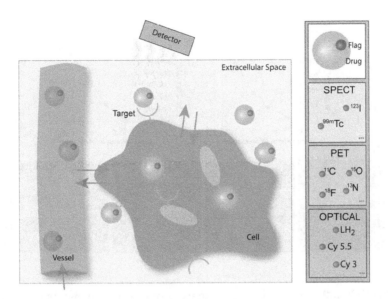

FIGURE 2.1: The tracer principle. Molecular targets are typically addressed and visualized by injection of tracers, which are consisting of a drug (light green) to which a flag (dark green; radioactivity, fluorescent dyes, quenched optical dyes, etc.) is attached. The tracer arrives in organs via blood vessels, diffuses into the extracellular space and can bind to externalized targets (orange; receptors, etc.) on the cell surface or cross the cell membrane to bind to intracellular targets. By its flag the tracer emits light or gamma rays to be detected from outside the organism by SPECT, PET or optical imaging. Examples of isotopes and dyes used for SPECT, PET and optical imaging are listed.

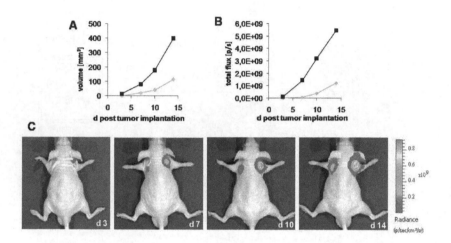

FIGURE 2.2: Longitudinal BLI analysis of the growth of subcutaneous brain tumors. CD1 nu/nu mice were injected twice with $1*10^6$ human U87ΔEGFR glioma cells that stably express firefly luciferase. At different days post tumor implantation the tumor volume was measured with the help of a caliper (A); tumor volume = $0.52 \times$ length $\times$ width^2. Furthermore, tumor activity was assessed by BLI measurements 10 min after intraperitoneal injection of 2 mg D-luciferin using the IVIS spectrum system (Caliper Life Sciences). Serial BL images displayed in pseudocolors superimposed to a white light image show tumor growth of two subcutaneous tumors at the left and right back region (C). BLI signals that arise from the tumors were quantified (B) (gray: left tumor, black: right tumor) to assess viable tumor volumes.

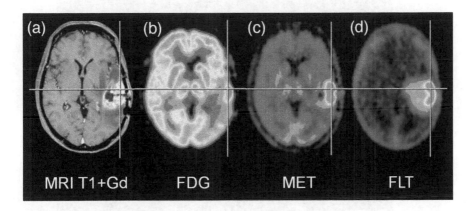

FIGURE 2.6: MRI and PET are being used together to assess (a) the breakdown of the blood-brain barrier (T1+Gd); (b) the metabolic activity of the tumor as assessed by [^{18}F]FDG-PET, which also serves as a surrogate marker for cellular density; (c) the uptake of radiolabelled amino acids such as [^{11}C]MET, which serves as direct marker for the expression of amino acid transporters and as a surrogate marker for neovascularization; (d) the uptake of radiolabelled thymidine ([^{18}F]FLT), which serves as direct marker for cellular thymidine kinase activity and as surrogate marker for cell proliferation. The various imaging markers give complementary information on the activity and extent of the tumor.

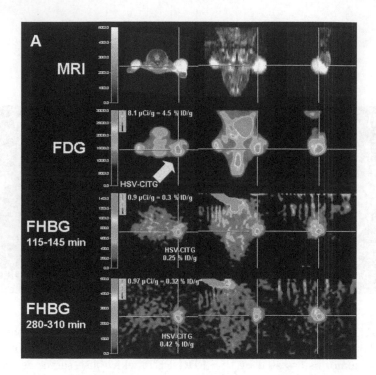

FIGURE 2.7: Imaging-guided gene therapy paradigm of experimental gliomas. Protocol for identification of viable target tissue and assessment of vector-mediated gene expression *in vivo* in a mouse model with three subcutaneous growing gliomas. Row 1: localization of tumors by MRI. Row 2: the viable target tissue as depicted by [18F]FDG-PET. Note the signs of necrosis in the lateral portion of the left-sided tumor (arrow). Rows 3–4: following vector-application into the medial viable portion of the tumor (arrow) the "tissue-dose" of vector-mediated gene expression is quantified by [18F]FHBG-PET. Row 3 shows an image acquired early after tracer injection, which is used for coregistration. Row 4 displays a late image with specific tracer accumulation in the tumor that is used for quantification.

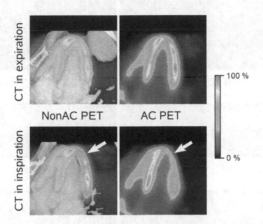

FIGURE 5.16: Cardiac FDG PET/CT scan (long axis view). Non-corrected PET images showing position of the heart during CT (top: end-expiration; bottom: end-inspiration) and PET are shown on the left side, attenuation-corrected PET images on the right. Note the apparent uptake defect in the lateral wall (arrow) due to misregistration between end-expiration CT and PET.

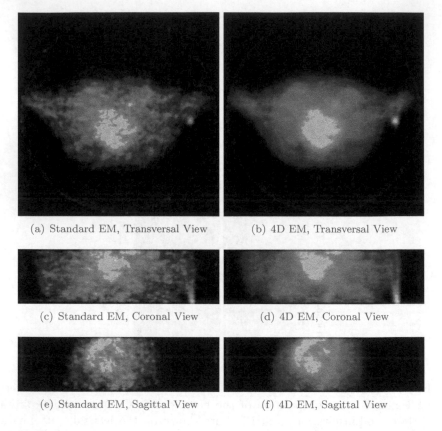

(a) Standard EM, Transversal View   (b) 4D EM, Transversal View

(c) Standard EM, Coronal View   (d) 4D EM, Coronal View

(e) Standard EM, Sagittal View   (f) 4D EM, Sagittal View

FIGURE 9.2: Two exemplary $H_2{}^{15}O$ reconstructions, acquired with the standalone Siemens PET Scanner ECAT EXACT (model 921). It should be noticed that with this type of scanner image acquisition can be done only slice-by-slice. The images on the left-hand side show standard EM reconstructions of a slice of the seventh frame in transversal, coronal and sagittal view. The reconstructions have been smoothed by additional Gauss filtering between each EM iteration. The images on the right-hand side show the same views of the same slice but computed via the described 4D-EM reconstruction method. Again, after each EM iteration Gauss filtering has been applied to ensure spatial smoothness. The images on the right-hand side have a much higher signal-to-noise ratio than those derived from the frame-independent standard EM reconstructions.

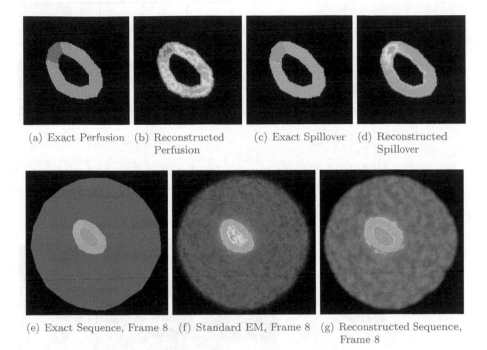

(a) Exact Perfusion (b) Reconstructed Perfusion (c) Exact Spillover (d) Reconstructed Spillover

(e) Exact Sequence, Frame 8 (f) Standard EM, Frame 8 (g) Reconstructed Sequence, Frame 8

FIGURE 9.3: A combined reconstruction/parameter identification process of a synthetic myocardial perfusion example. Figure 9.3(a) and Figure 9.3(b) show the exact perfusion and the reconstruction with the method proposed in [3]. Figure 9.3(c) and Figure 9.3(d) show the exact arterial spillover and its reconstruction. Figure 9.3(e)–Figure 9.3(g) show the 8th frame of the underlying exact image sequence, the frame-independent standard EM reconstruction and the reconstruction method proposed in [3].

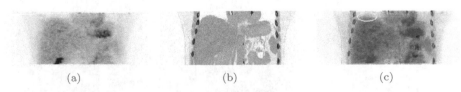

(a)   (b)   (c)

FIGURE 9.5: Representation of one temporal bin of the acquisition gated with the respiration: (a) gated PET image non corrected for attenuation, (b) gated CT image, (c) overlaid image of both PET and CT gated images. A misalignment of 6/7mm between the PET and CT images can be seen at the level of the diaphragm on the overlaid image inside the yellow circle.

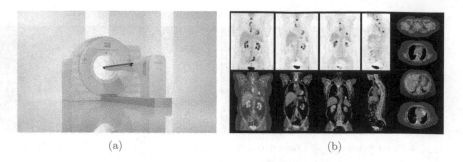

FIGURE 10.1: PET/CT. (a) Siemens Biograph® mCT. (b) 5-min ultraHD PET study of an obese patient with lung CA (data courtesy of University of Tennessee, Knoxville, TN, USA).

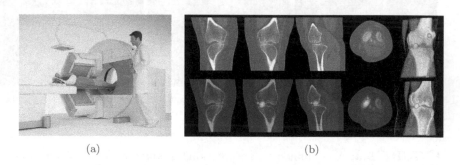

FIGURE 10.2: SPECT/CT. (a) Siemens Symbia® T16 True-point™SPECT/CT. (b) SPECT/CT delineation of subchondral cyst in left knee joint (data courtesy of PRP Cumberland Diagnostic Imaging, New South Wales, Australia).

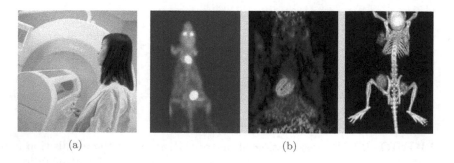

FIGURE 10.3: PET-SPECT-CT. (a) Siemens Inveon® MultiModality. (b) Preclinical studies (courtesy of University of Wisconsin, Madison, WI and Eberhard Karls University Tuebingen, Tuebingen, Germany).

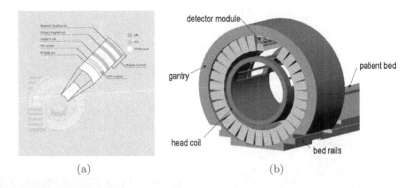

(a)   (b)

FIGURE 10.4: MR-PET: (a) detector layout of Siemens Brain-PET™(prototype) and (b) design study of whole-body MR-PET.

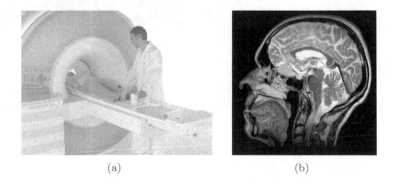

(a)   (b)

FIGURE 10.5: MR-PET: Prototype. (a) Siemens prototype BrainPET™ (Works in Progress, The product is under development and is not commercially available. (b) MR-PET study (courtesy of University Tübingen, Germany).

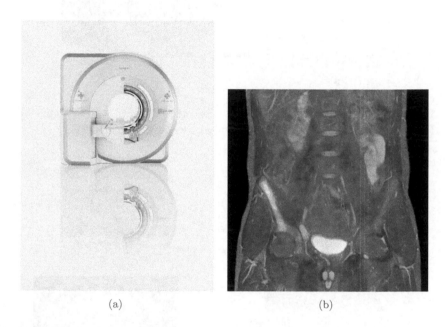

(a)   (b)

FIGURE 10.6: Fully integrated whole-body MR-PET. (a) Siemens Bio-
graph™mMR; the Biograph™mMR system requires 510(k) review by the FDA
and is not commercially available. Due to regulatory reasons its future availability in any country cannot be guaranteed. Please contact your local Siemens
organization for further details. (b) Simultaneously acquired MR-PET showing bone marrow imaging in case of cancer.

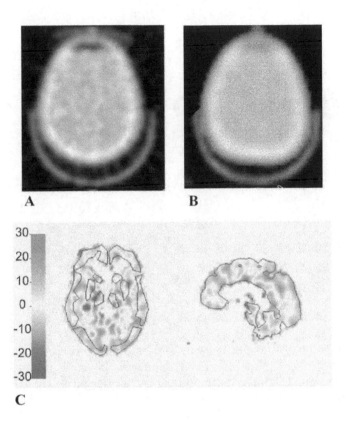

FIGURE 11.4: MR-AC for brain PET: atlas-based attenuation correction. (A) Attenuation map measured through a PET transmission scan; (B) Attenuation map obtained through atlas registration and addition of the head holder; (C) Coronal and saggital view of voxel-by-voxel relative differences between PET attenuation corrected using attenuation maps A and B. (From [33].)

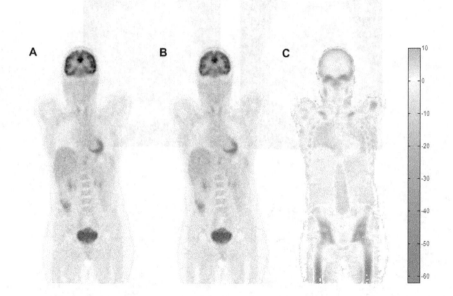

FIGURE 11.9: Effect of ignoring cortical bone during AC. (A) PET image reconstructed using the original CT image. (B) PET image reconstructed using the same CT image with all bone structures set to the HU value of soft tissue thus simulating a best case scenario of MR-AC where bone attenuation is ignored. (C) Relative difference (%) between (A) and (B) illustrating the largest effect inside the skeleton. Note, voxels set to white in low uptake regions with SUV<0.2 in the original PET image.

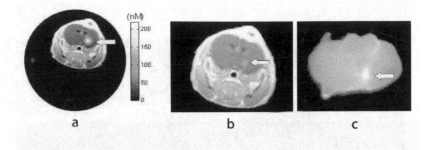

a   b   c

FIGURE 12.7: Combined FMT and MRI imaging. Examples of fluorescence imaging using a cathepsin B-activatable imaging probe. (a) and (b), Enzyme activity in a 9L glioma model in a live mouse. The image in (a) is superimposed onto an MRI image shown separately in (b) with gadolinium enhancement of the glioma28. (c), In vitro FRI of the axial brain section corresponding to the MR and FMT images. The tumor position is indicated by the arrow. (From [29].)

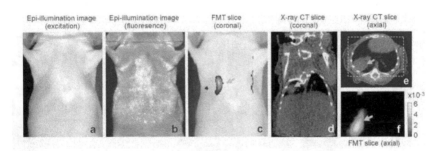

FIGURE 12.8: Tomographic imaging of fluorescent proteins and corresponding X-ray CT from a nude mouse implanted with GFP-expressing lung tumors, obtained 10 days post-image implantation. (a) Epi-illumination image of the mouse at the excitation wavelength; (b) Epi-illumination image at the emission wavelength showing high skin autofluorescence. (c) Tomographic slice (in color, after threshold was applied) obtained from the tumor depth (7 mm from top surface) overlaid on the white light image of the mouse. (d, e) CT coronal and axial slices, respectively; the tumor position is marked by arrows. (f) Axially reconstructed slice corresponding to the yellow dashed rectangle on (e). (From [31].)

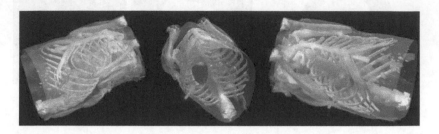

FIGURE 12.9: Multimodality imaging. The fluorescent reconstructions of Figure 12.8 rendered simultaneously with X-ray CT images. The tumor is indicated by an arrow. (From [31].)

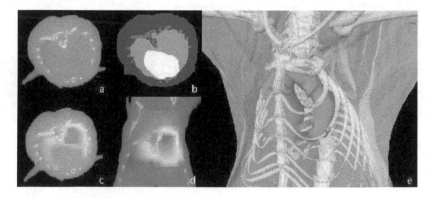

FIGURE 12.11: Reconstruction based on data from combined FMT-XCT setup. (a) X-ray slice. (b) Segmentation of X-ray data in lungs, heart, bone and remaining tissue. (c) Reconstruction of fluorescent biodistribution in lung, transversal slice and (d) sagittal slice. (e) 3D-hybrid visualization. (From [22].)

Chapter 7

Image Processing Techniques in Emission Tomography

Fabian Gigengack, Michael Fieseler, Daniel Tenbrinck, and Xiaoyi Jiang

Department of Mathematics and Computer Science, University of Münster, Münster, Germany

7.1	Introduction		119
7.2	Denoising		121
	7.2.1	Image domain	122
	7.2.2	Fourier transform domain	123
	7.2.3	Wavelet transform domain	124
7.3	Interpolation		126
7.4	Registration		129
	7.4.1	Categorization	130
		7.4.1.1 Nature of transformation	132
		7.4.1.2 Similarity measure	133
	7.4.2	Validation	135
	7.4.3	Software	137
7.5	Partial volume correction		137
	7.5.1	The partial volume effect in PET imaging	138
	7.5.2	Correction methods	140
7.6	Super-resolution		144
7.7	Validation		146
	7.7.1	Intensity-based measures	146
	7.7.2	Phantoms	148
		7.7.2.1 Hardware	148
		7.7.2.2 Software	149
References			150

7.1 Introduction

Image processing is an essential part of medical imaging in general, and emission tomography in particular. Formally, an output image (or multiple images or an image description) with desired properties is generated by processing one or more input images. Many common image processing techniques from computer science are used in emission tomography, e.g., noise reduction or image registration for motion correction.

The transfer of general image processing techniques to medical imaging is usually straightforward but sometimes specific modifications are necessary or beneficial. For instance, medical imaging is usually associated with 3D volumes or even 4D time series, unlike 2D images in traditional image processing. Additionally, the inclusion of prior knowledge, e.g., information about anatomy or the acquisition process, can be advantageous.

The fields of application are manifold—and so are the image processing algorithms. Accordingly, we can give only a non-exhaustive overview of basic techniques commonly used in emission tomography. This chapter is mainly aimed at those not familiar with image processing and may serve as a starting point for further studies.

Noise removal is an important preprocessing step for many tasks in image processing—especially in medical image analysis. In Section 7.2 we describe common methods for noise removal and distinguish between image, Fourier, and wavelet transform domain based methods. By definition, image processing denotes the manipulation of images. This manipulation often includes resampling of the data, i.e., interpolation, as discussed in Section 7.3. Image registration provides the fundamentals for image alignment and motion detection. Hence, we present a categorization of registration techniques in Section 7.4 where we put a particular focus on the meaning of "similarity." We conclude the section with a brief review of current registration software packages. When approaching the limitations of scanner resolution, the partial volume effect (PVE) becomes relevant. An introduction to PVE and techniques dealing with the correction of PVE are presented in Section 7.5. Super-resolution is a promising approach to overcome resolution limitations of image acquisition and is discussed in Section 7.6. A major problem in developing algorithms in medical imaging is the lack of ground-truth data. This applies to a wide range of algorithms, e.g., motion detection and segmentation. We discuss general validation methods including soft- and hardware phantoms in Section 7.7.

All figures in this chapter were created using MATLAB®. For Wiener filtering, Fourier and wavelet transform, and deconvolution the respective functions provided by MATLAB® were used. PET images are shown using an inverted grayscale colormap so that dark colors indicate high activity and light colors low activity, respectively.

7.2 Denoising

A variety of intrinsic factors degrade image quality in emission tomography. These include scatter, randoms, out-of-field counts, detector dead time, detector noise, patient motion, attenuation, non-colinearity of photons, positron range and image reconstruction artifacts. Further, a trade-off between image quality on the one hand and examination time and radiation exposure on the other is required. Prolonged examination time usually implies inconvenience for the patient. In contrast, short acquisition time and low radiation dose lead to a reduction of image quality due to a lowered statistic.

In the following, we describe techniques for noise removal. The focus is put on denoising as a post-reconstruction process, leaving aside noise reduction during image reconstruction. The image degradation process can be modeled as

$$I = P \star I_u + N \, , \tag{7.1}$$

where I is the measured image signal, I_u is the (uncorrupted) image free of noise, P a convolution mask and N additive noise. The discrete convolution of an $n \times m$ mask P and an image I_u at pixel (x, y) is defined as

$$(P \star I_u)(x, y) := \sum_{i=1}^{n} \sum_{j=1}^{m} P(i, j) I_u(x - i, y - j) \, . \tag{7.2}$$

The focus of this section is put on the elimination of the additive noise N. Making good estimates of P is an important part of the more general problem of image restoration. Image restoration is discussed in connection with partial volume correction (PVC) in Section 7.5 where P represents the point spread function (PSF).

Compared to the underlying image I_u, the additive noise N is mainly present in high frequencies. In order to improve image quality by eliminating the high-frequency noise component the image has to be low-pass filtered. Image quality can be quantified by means of the signal-to-noise ratio (SNR). The definition of the SNR is given in Section 7.7.1.

As denoising has a long history, many approaches exist. A short introduction into basic theories is given in the following. For further reading a survey of denoising techniques is given in [45]. Publications dealing especially with denoising in emission tomography are [1, 6, 24, 59]. Standard image processing literature usually deals with this topic as well [21, 22, 48, 52].

The rest of this section is structured according to three main approaches for image denoising. First, image domain–based filter methods are discussed in Section 7.2.1. They can be understood intuitively and thus provide a good starting point. In Section 7.2.2 filtering in the frequency domain using the Fourier transform is presented. Finally, the actual popular filtering in the wavelet transform domain is introduced in Section 7.2.3.

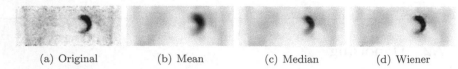

| (a) Original | (b) Mean | (c) Median | (d) Wiener |

FIGURE 7.1: The original noisy image (a) of a human heart was filtered using different standard denoising techniques. (b) Mean filtering. (c) Median filtering. (d) Wiener filtering.

7.2.1   Image domain

To get a sense of denoising it is helpful to look at basic image domain–based filtering techniques. In the following, denoising is illustrated using 2D examples of *mean filtering*, *median filtering*, and a local adaptive version of *Wiener filtering*.

An example of denoising in PET is given in Figure 7.1. The original image is shown in Figure 7.1(a). The noise visible in the image is caused by low statistics. The amount of noise is reduced using mean filtering in Figure 7.1(b), median filtering in Figure 7.1(c), and Wiener filtering in Figure 7.1(d). In the following these techniques are discussed in detail.

Mean filtering
The mean filter is a linear filter technique. Each pixel of the image is replaced by the mean value of its neighborhood. Technically, the noisy input image I is convolved with a filter mask M delivering the denoised image

$$I_d = M \star I. \tag{7.3}$$

For a filter size of 3×3 the mean filter mask is defined as

$$M := \frac{1}{9} \begin{pmatrix} 1 & 1 & 1 \\ 1 & 1 & 1 \\ 1 & 1 & 1 \end{pmatrix}. \tag{7.4}$$

M is always scaled in such a way that its values sum up to 1. This normalization prevents a shift of intensities. Mean filtering leads to blurring and does not preserve edges as is apparent in Figure 7.1(b).

A related approach is *Gaussian filtering* where the filter mask M is defined as a (two-dimensional) Gaussian function.

Median filtering
Median filtering is a non-linear method which is in particular suited to remove salt and pepper noise. For every pixel of the image a neighborhood is chosen which is, including the pixel itself, sorted by value. The median of these sorted values is chosen as the new value for the pixel. Looking at the torso outline in Figure 7.1(c) it can be seen that edges are preserved better compared to mean filtering.

Wiener filtering

The Wiener filter was introduced in the 1940s by Norbert Wiener. It is an optimal filter in the sense of minimizing the mean-squared error (MSE) of the denoised image I_d and the true image I_u. In the following, a local adaptive version of the Wiener filter, as described in [39], is discussed. For each pixel (x, y), local statistical characteristics such as the mean $\mu(x, y)$ and the variance $\sigma^2(x, y)$ are estimated from a local neighborhood and are used to calculate the new pixel value $I_d(x, y)$:

$$I_d(x, y) = \frac{\sigma_f^2(x, y)}{\sigma_f^2(x, y) + \sigma_{avg}^2} \left(I(x, y) - \mu(x, y) \right) + \mu(x, y) , \qquad (7.5)$$

where σ_{avg}^2 denotes the average of all local variances $\sigma(\cdot, \cdot)$ and serves as a variance estimation of the additive noise N of I. Further $\sigma_f^2(x, y)$ is defined as

$$\sigma_f^2(x, y) = \begin{cases} \sigma^2(x, y) - \sigma_{avg}^2 , & \text{if } \sigma^2(x, y) > \sigma_{avg}^2 \\ 0 & \text{otherwise} . \end{cases} \qquad (7.6)$$

In Figure 7.1(d) the effect of the adaptive character of the local Wiener filter can be clearly seen in the area around the heart. The noise in the background is filtered strongly whereas the region of the heart with high uptake is filtered to a lesser extent.

7.2.2 Fourier transform domain

Contrary to the image domain, images can be represented by their inherent frequencies in the Fourier transform domain. As noise is primarily present in high frequencies this is a suitable representation to eliminate noise efficiently.

In Fourier domain based filtering the noisy image I is transformed to frequency space using $\mathcal{F}$, the (fast) Fourier transform (FFT). A filter mask $M_{\mathcal{F}}$ is developed in frequency space and applied to the transformed image $\mathcal{F}(I)$. Subsequently, the filtered frequency image is transformed back to image space using the inverse FFT $\mathcal{F}^{-1}$:

$$I_d = \mathcal{F}^{-1}(M_{\mathcal{F}} \cdot \mathcal{F}(I)) , \qquad (7.7)$$

where the multiplication is applied point-wise. As $\mathcal{F}^{-1}$ can lead to complex values, the real part of the inverse FFT should be taken here. Obviously, the filtering or denoising result depends on the choice of an adequate filter mask $M_{\mathcal{F}}$.

We depict the proceeding of Fourier domain filtering with a frequency-based realization of the mean filter, as illustrated in Figure 7.2. Equation (7.3) together with the convolution theorem (C.T.) leads to

$$I_d = M \star I = \mathcal{F}^{-1}(\mathcal{F}(M \star I)) \overset{C.T.}{=} \mathcal{F}^{-1}(\mathcal{F}(M) \cdot \mathcal{F}(I)) , \qquad (7.8)$$

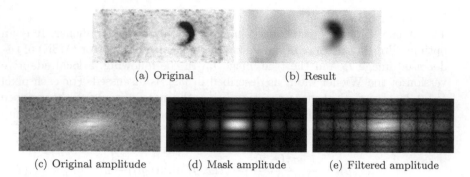

(a) Original   (b) Result

(c) Original amplitude   (d) Mask amplitude   (e) Filtered amplitude

FIGURE 7.2: The original noisy image (a) was filtered in Fourier domain using the frequency version of the mean filter. (b) Mean filtering result. (c) Amplitude of the original image $|\mathcal{F}(I)|$. (d) Amplitude of the filter mask $|\mathcal{F}(M)|$. (e) Amplitude of $|\mathcal{F}(M) \cdot \mathcal{F}(I)|$.

where M is the mean filter mask. Hence, convolution in the image domain is equivalent to a point-wise multiplication in frequency domain.

The amplitude of the FFT of the noisy input image $\mathcal{F}(I)$ is shown in Figure 7.2(c). The amplitude of the Fourier-transformed mean filter $\mathcal{F}(M)$ can be seen in Figure 7.2(d). The result of $\mathcal{F}(M) \cdot \mathcal{F}(I)$ is illustrated in Figure 7.2(e). The resulting filtered image I_d in Figure 7.2(b) is equal to the result in Figure 7.1(b) (up to machine precision).

One of the advantages of filtering in the Fourier domain is the possibility of performing deconvolution, as described in Section 7.5.2. A popular deconvolution filter is the frequency space version of the Wiener filter. The characteristics lie in the special choice of the deconvolution mask $M_\mathcal{F}$. For a detailed discussion regarding this filter we refer to [21, 22].

7.2.3   Wavelet transform domain

Similar to the Fourier transform, wavelets represent images as a linear combination of basis functions of different frequencies. Wavelets have local support in image space, whereas Fourier basis functions have infinite support. Thus, a disadvantage of Fourier analysis is the total loss of local information as compared to the wavelet transform.

Wavelet-based filters are very promising at present. There is a huge number of publications on this topic. A brief overview is given in [45]. Wavelet filters can be subdivided into linear and nonlinear methods. A wavelet realization of the Wiener filter is a typical example of linear filters. The key principle of non-linear procedures is the property that white noise in the image domain maps to white noise in the transformed domain; thus noise can be separated efficiently from the pure signal.

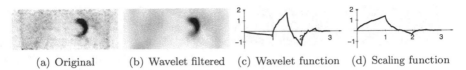

(a) Original   (b) Wavelet filtered   (c) Wavelet function   (d) Scaling function

FIGURE 7.3: The original noisy image (a) was filtered using Daubechies wavelets (D4). (b) Wavelet filtering result. (c) D4 wavelet function. (d) D4 scaling function.

A wavelet transform is performed for a finite number of levels. On each level the image is processed using a *wavelet function* which is in principle a band-pass filter (for high frequencies). The wavelet function is scaled on each level, i.e., its bandwidth is halved compared to the previous level. The filter results of the differently scaled wavelet functions as a whole are called *details*. To cover the whole frequency spectrum a *scaling function*, in effect a low-pass filter, preserves low frequencies. The filter result of the scaling function is called *approximation*. The original image can be recovered completely from the wavelet representation of details and approximation.

In general, non-linear wavelet denoising consists of three steps:

1. Transformation of the original image into the wavelet transform domain.

2. Thresholding of the details (on each level).

3. Transformation back into the image space.

Apart from choosing suitable basis functions, the main problem is to determine an adequate threshold value. Adaptive as well as non-adaptive methods for threshold determination have been proposed [45].

An illustrative example can be found in Figure 7.3. The noisy image in Figure 7.3(a) was filtered with *D4 Daubechies wavelets* on the basis of the three steps described above with a manually chosen threshold. The D4 basis functions are given in Figure 7.3(c) and 7.3(d).

According to the three steps of wavelet denoising, some intermediate results of the 3-level D4 wavelet transformation are shown in Figure 7.4. The details are shown as examples in Figures 7.4(a) and 7.4(b) for level 1 and 3, respectively. After applying thresholding to the details, *reduced* versions result which are shown in Figures 7.4(c) and 7.4(d). The next step is the transformation back into the image domain yielding the image shown in Figure 7.3(b).

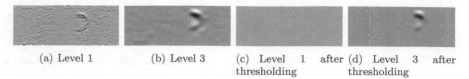

| (a) Level 1 | (b) Level 3 | (c) Level 1 after thresholding | (d) Level 3 after thresholding |

FIGURE 7.4: Some detail images of the 3-level D4 wavelet transform from Figure 7.3 before and after thresholding. (a) Level 1 details before thresholding. (b) Level 3 details before thresholding. (c) Level 1 details after thresholding. (d) Level 3 details after thresholding.

7.3   Interpolation

Interpolation denotes the continuous approximation of a discrete signal. This is required for many tasks in medical imaging in general and emission tomography in particular. For example, registration and visualization techniques require interpolation as discrete images need to be transformed (e.g., rotation, translation, scaling) and hence resampled at an irregular grid. We will introduce the basic concept of interpolation on the basis of standard interpolation schemes: *sinc*, *nearest neighbor* (NN), *linear*, and *spline* interpolation. As the choice of an adequate interpolation method depends on the task-specific requirements, we refer to [36, 49] for a more detailed discussion.

According to the sampling theorem, a continuous image can be reconstructed exactly from its uniformly spaced samples if the highest frequency present in the image is lower than the Nyquist frequency, i.e., half the sampling rate. A convolution with the sinc function in the image domain represents such an ideal interpolation.

Definition 1 (Sinc)   *The one-dimensional* sinc *function is defined as*

$$\text{Sinc}(x) := \begin{cases} \frac{\sin(\pi x)}{\pi x}, & x \neq 0 \\ 1, & x = 0 \, . \end{cases} \tag{7.9}$$

A disadvantage of the sinc function is that it is spatially unlimited and thus improper for practical use in the image domain. So the task is to find spatially limited interpolation functions with characteristics similar to the sinc function. These characteristics are discussed in [36] on the basis of spatial and Fourier analysis, computational complexity as well as runtime evaluations, and determination of qualitative and quantitative interpolation errors.

The one-dimensional NN, linear, and spline interpolation are defined in the following. The extension to higher dimensions is straightforward in each case. For a more detailed description of spline interpolation we refer to [43], as FAIR was used for all spline interpolations in this section.

Definition 2 (NN) *The one-dimensional* nearest neighbor *interpolation of a discrete image I at position x is defined as*

$$NN(I, x) := I([x]) \,, \tag{7.10}$$

where $[x]$ is the rounded value of x.

Definition 3 (Linear) *The one-dimensional* linear *interpolation of a discrete image I at position x is defined as*

$$Linear(I, x) := (1 - \alpha_x)I(\lfloor x \rfloor) + \alpha_x I(\lceil x \rceil) \,, \tag{7.11}$$

where $\alpha_x = x - \lfloor x \rfloor$. $\lfloor \cdot \rfloor$ and $\lceil \cdot \rceil$ are the floor and ceiling functions.

Definition 4 (Spline) *The one-dimensional* spline *interpolation of a discrete image I is defined as the function minimizing the bending energy*

$$\mathcal{E}(I) = \int_\Omega (\mathrm{Spline}(I, x)'')^2 dx \tag{7.12}$$

subject to

$$\mathrm{Spline}(I, i) = I(i), \quad \forall i \in \Omega \cap \mathbb{N} \,. \tag{7.13}$$

The NN and linear interpolation as defined above can also be interpreted as a convolution of the discrete signal with an accordant interpolation function. The appropriate filter function for the sinc, NN, and linear interpolation are plotted in Figure 7.5 for $x \in [-3, 3]$. These plots illustrate that the linear interpolation function is a better approximation of the ideal sinc function compared to NN.

A comparison of the described interpolation schemes is illustrated for a 1D example in Figure 7.6. The given data is indicated by the five big dots. The solid lines show the nearest neighbor interpolation with unwanted jump discontinuities. Linear interpolation is shown with a dashed line. The dotted line represents the smooth spline interpolation. As a reference, the ideal sinc interpolation is plotted with a dash-dot line.

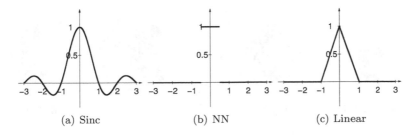

(a) Sinc   (b) NN   (c) Linear

FIGURE 7.5: Interpolation functions for $x \in [-3, 3]$. (a) Sinc interpolation function. (b) NN interpolation function. (c) Linear interpolation function.

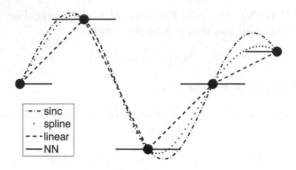

FIGURE 7.6: Five data points (bold dots) are interpolated with *sinc* (dash-dot line), *spline* (dotted line), *linear* (dashed line), and *nearest neighbor* (solid line) interpolation.

This illustrative one-dimensional example already points out some similarities and differences of different interpolation mechanisms. When choosing an interpolation method there are different criteria to be considered such as computation time, differentiability, continuity, and accuracy. Among the presented methods, nearest neighbor interpolation is the fastest one but it lacks meaningful derivatives and the results are (in general) not continuous. A differentiable and continuous method is spline interpolation, but at the expense of higher computation costs. Linear interpolation is a compromise as it is relatively fast and continuous. Differentiability is given, except for regular grid points.

The impact of interpolation is illustrated in Figure 7.7 by scaling up real PET data. The original image in Figure 7.7(a) shows a human heart. This image was scaled up by a factor of 20 using *NN* (Figure 7.7(b)), *linear* (Figure 7.7(c)), and *spline* (Figure 7.7(d)) interpolation. It is obvious that linear and spline interpolation are superior to NN. The differences between linear and spline interpolation are not that clear at first glance.

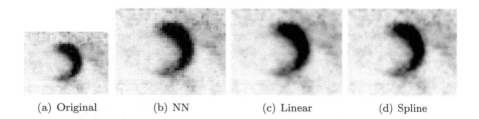

(a) Original   (b) NN   (c) Linear   (d) Spline

FIGURE 7.7: The original image (a) was scaled up using (b) NN, (c) linear, and (d) spline interpolation.

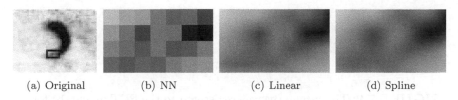

| (a) Original | (b) NN | (c) Linear | (d) Spline |

FIGURE 7.8: Detail of the scaling example from Figure 7.7. (a) Detail location indicated by black rectangle. Interpolation results for detail region using (b) NN, (c) linear, and (d) spline interpolation.

To determine the advantages of spline interpolation we look at a detail in Figure 7.8(a) indicated by the rectangle. The inferiority of NN interpolation gets even more evident. Furthermore, the spline interpolation looks smoother compared to the linear interpolation which is only piecewise differentiable. Many tasks in medical image processing need continuously differentiable images as input, e.g., minimization of the distance functional for registration (see Equation (7.15)). Hence, from a theoretical point of view it is essential to use spline interpolation for such tasks as the differentiability is given. Yet, fast and simple methods like linear interpolation are often sufficient in practice.

Apart from the standard interpolation methods discussed in this section a large variety of algorithms exists. A diffusion-based interpolation technique with the intention to reduce the partial volume effect (see Section 7.5) was presented in [53]. In [68] a PDE-based framework for interpolation and regularization of scalar- and tensor-valued images is presented that does not require a regular grid. The method based on anisotropic diffusion allows discontinuity-preserving interpolation without additional oscillations.

7.4   Registration

Correction techniques in emission tomography images often include *registration*. Registration denotes the spatial alignment of two corresponding images. In medical imaging these corresponding images are obtained from the same or different patients, acquired at the same or at a different time, using the same or different scanning techniques. One image is transformed in order to match the other image as well as possible.

This simple definition raises some fundamental questions: How can the *similarity* of two images be measured in order to find a good match? How can the *optimal spatial transformation*, maximizing the similarity, be found?

Before answering these questions, an introductory example of registration in emission tomography is discussed (Figure 7.9). Attenuation correction in

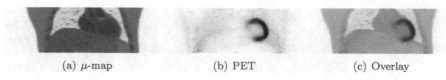

(a) μ-map   (b) PET   (c) Overlay

FIGURE 7.9: The μ-map (a) was registered to the PET image (b) for attenuation correction. (c) Overlay of the PET image and the μ-map. The colormap for the μ-map is chosen similarly to the colormap of the PET data (inverted grayscale) for better visibility.

PET is primarily the correction for absorption of photons in the body. This is done by registering an attenuation map (μ-map) to the PET data and incorporating this registered attenuation map into a second reconstruction of the PET data. The μ-map is usually obtained from a separate CT scan. In Figure 7.9(a) the μ-map registered to the PET image in Figure 7.9(b) is shown. An overlay of the μ-map and the PET image is shown in Figure 7.9(c) to visualize the accuracy of the alignment.

As seen in this introductory example, registration in emission tomography is inevitable due to several tasks requiring spatial alignment of corresponding images. Especially in combined approaches like PET/CT or PET/MRI an optimal alignment is mandatory as information of both techniques (functional and morphological) is combined. Also, there are numerous examples of monomodal registration tasks for scans taken at different times (e.g., tumor growth studies, motion correction in gated PET (see Chapter 8), rest-stress comparisons) or of different objects (e.g., normalization to atlas data).

In the following, image registration algorithms are categorized in Section 7.4.1. In Section 7.4.2 methods for validating registration results are discussed. For practical use some freely available registration software is presented in Section 7.4.3. For sustained discussion beyond the scope of this general overview we refer to some comprehensive survey articles about registration in general [4, 9, 13, 72] and medical imaging in particular [23, 27, 37, 41, 55, 61].

7.4.1 Categorization

Throughout the rest of this section the following notation will be used. The image to be transformed is the *template image* $\mathcal{T}$ and the static image is the *reference image* $\mathcal{R}$. Let d denote the spatial dimension and $\Omega \subset \mathbb{R}^d$ the domain. Then the *template image* $\mathcal{T}$ and the *reference image* $\mathcal{R}$ are defined as $\mathcal{T}, \mathcal{R} : \Omega \to \mathbb{R}$.

To get an overview of the huge variety of different registration approaches, a categorization using criteria adopted from [41] follows. In this context the nature of transformations and similarity measures are discussed in more detail in Sections 7.4.1.1 and 7.4.1.2.

Dimensionality
The spatial dimension of most data in emission tomography is 3D. Nevertheless, various 2D/2D and 2D/3D cases exist where single slices or projections are aligned to other slices or 3D volumes [54].

Nature of registration basis
A general distinction of registration techniques can be made by the nature of the registration basis. Some registration techniques use intrinsic or extrinsic landmarks [17, 60] while others are segmentation based [12]. These methods play no major role in emission tomography. The most essential class of registration methods is voxel-intensity based.

User interaction
Some registration techniques require user interaction while others are fully automated. User interaction can vary between complete manual alignment or delineation of sparse landmarks. Fully interactive methods are usually accompanied by sophisticated visualization techniques [38]. A rough manual initialization or visual validation of the registration results can be understood as user interaction as well. The amount of user interaction needed usually depends on the following criteria: accuracy, reliability, computational costs and convenience. User interaction can lead to more accurate and reliable results with relatively low computational costs at the expense of inconvenience.

Modality
The main modalities in emission tomography are PET and SPECT. As PET and SPECT images provide functional information, they are often combined with supplemental morphological scanning techniques to increase the information content [38, 57, 60]. These additional modalities are mainly CT, MRI and ultrasound. An example for PET/CT to μ-map– and PET to μ-map–registration is given in Figure 7.9.

Subject
If only one patient is involved in the registration task it is called *intrasubject*. When two datasets of different patients are registered it is denoted *intersubject*. A third case is the alignment of patient and *atlas* data.

Object
Many registration papers in medical imaging refer to the head [33]. As the head is a rigid object, the transformations in monomodal registration problems are usually restricted to rigid ones. But especially in the field of emission tomography there are many additional applications of interest like the detection of coronary artery disease. Registration is a helpful instrument to support examinations of the heart, e.g., by eliminating cardiac [20, 15] and respiratory motion (see Chapter 8). Additionally, tumor studies are of interest in emission tomography, requiring registration as well [69].

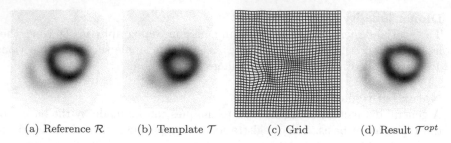

(a) Reference $\mathcal{R}$ (b) Template $\mathcal{T}$ (c) Grid   (d) Result $\mathcal{T}^{opt}$

FIGURE 7.10: The template image $\mathcal{T}$ in (b) is registered non-linearly to the reference image $\mathcal{R}$ in (a) using FAIR [43]. The resulting transformation is expressed by the grid in (c). $\mathcal{T}$ is resampled at the grid points yielding the result $\mathcal{T}^{opt}$ in (d).

Figure 7.10 gives an example of registration in cardiac-gated PET of a mouse. The template image $\mathcal{T}$ showing the heart during the systole is registered to the reference image $\mathcal{R}$ showing the heart during the diastole using a non-parametric transformation model. The similarity is measured with the sum of squared differences (see Definition 5). The heart contraction during the cardiac cycle can be clearly seen in these images. The task is to find an optimal transformation (grid) such that the warped image $\mathcal{T}^{opt}$ is as similar to $\mathcal{R}$ as possible. $\mathcal{T}^{opt}$ results from interpolating $\mathcal{T}$ (see Section 7.3) at the irregular transformation grid.

7.4.1.1   Nature of transformation

Transformations can be roughly divided into *parametric* and *non-parametric* ones. While parametric transformations cover rigid, affine or spline (free-form) deformations, non-parametric transformations denote deformations that are independent for each voxel. Rigid and affine models are appropriate mainly for intrasubject registration tasks while non-parametric models apply better to intersubject or atlas registration. Nevertheless, there are also many cases where non-parametric (parametric) models are used in intrasubject (intersubject) studies.

In most registration setups the transformation applies to the whole image domain. These setups are referred to as *global* approaches. But in some cases the transformation only applies to *local* parts of the image [57].

Parametric registration methods
In parametric image registration the transformation is given in terms of a parametric function, which is defined by a certain number of parameters. For example, the 3D rigid transformation parameter vector has six entries: three for describing the rotation around the three coordinate axes and three for translation. Hence, the task in parametric registration is to find the parameter

set p minimizing the distance between $\mathcal{R}$ and the transformed input image $\mathcal{T}_p$:

$$\arg\min_p \; \mathbb{D}(\mathcal{T}_p, \mathcal{R}) \,, \tag{7.14}$$

where $\mathcal{T}_p$ is the template image $\mathcal{T}$ transformed with the transformation according to the parameter vector p and $\mathbb{D}$ is the distance according to some similarity measure (see Section 7.4.1.2).

The problem of finding the optimal parameter set has to be solved in an *optimization* step. Optimization is a very important part of the registration process. Information about popular non-linear least square methods like Gauss–Newton or Levenberg–Marquardt and quasi-Newton methods such as (L-)BFGS or SR1 are given in [47].

Additional regularization is useful for some parametric methods like free-form deformations. Regularization is commented on further at the end of the following paragraph.

Non-parametric registration methods
In contrast to parametric registration, the transformation in non-parametric registration is not represented by a relatively small number of parameters. Instead, a non-parametric transformation $t : \mathbb{R}^d \to \mathbb{R}^d$ describes the deformation for every spatial position independently. The variational formulation of the registration problem is:

$$\arg\min_t \; \mathbb{D}(\mathcal{T}_t, \mathcal{R}) + \alpha \, \mathbb{S}(t) \,. \tag{7.15}$$

Here $\mathcal{T}_t$ is the template image $\mathcal{T}$ transformed according to the transformation t and $\mathbb{D}$ measures the distance. $\mathbb{S}$ denotes the regularization of the transformation and $\alpha \in \mathbb{R}^{\geq 0}$ is a weighting factor.

As non-parametric image registration is ill-posed [18], regularization is essential to find reasonable transformations. Regularization restricts the space of possible transformations to a smaller set of reasonable functions, e.g., by penalizing non-smooth transformations. The most important regularizers are elastic, fluid, diffusion and curvature regularization. Elastic regularization is a common choice in medical imaging. More information about regularization and additional constraints can be found in [18, 43, 46].

7.4.1.2 Similarity measure

To take up the questions that came up at the very beginning of Section 7.4, the definition of similarity is a very important part of image registration. In Equations (7.14) and (7.15) the similarity measure is represented by the distance functional $\mathbb{D}$. Equivalently, it is possible to use dissimilarity for comparison of two images (see Definitions 6 and 7). As $\mathbb{D}$ has to be minimized, such dissimilarity measures require some slight modifications.

The similarity measure has to be chosen according to the nature of the data to be registered. In monomodal studies it is usually sufficient to use the fast and easy to implement sum of squared differences.

Definition 5 (SSD) *The* sum of squared differences (SSD) *of two images $\mathcal{T}$ and $\mathcal{R}$ is defined as*

$$\mathrm{SSD}(\mathcal{T}, \mathcal{R}) := \|\mathcal{T} - \mathcal{R}\|^2 = \int_\Omega (\mathcal{T}(x) - \mathcal{R}(x))^2 \, dx \, . \tag{7.16}$$

SSD measures the point-wise distances of image intensities. For images with locally similar intensities the measure gets low. SSD has to be minimized and has its optimal value at 0.

Another common measure for monomodal studies is the normalized cross-correlation.

Definition 6 (NCC) *The* normalized cross-correlation (NCC) *of two images $\mathcal{T}$ and $\mathcal{R}$ is defined as*

$$\mathrm{NCC}(\mathcal{T}, \mathcal{R}) := \frac{\langle \mathcal{T}_u, \mathcal{R}_u \rangle}{\|\mathcal{T}_u\| \, \|\mathcal{R}_u\|} = \frac{\int_\Omega \mathcal{T}_u(x) \mathcal{R}_u(x) \, dx}{\sqrt{\int_\Omega \mathcal{T}_u(x)^2 \, dx} \sqrt{\int_\Omega \mathcal{R}_u(x)^2 \, dx}} \, . \tag{7.17}$$

where $\mathcal{T}_u = \mathcal{T} - \mu(\mathcal{T})$ and $\mathcal{R}_u = \mathcal{R} - \mu(\mathcal{R})$ are the unbiased versions of $\mathcal{T}$ and $\mathcal{R}$. For an image $\mathcal{I}$, $\mu(\mathcal{I})$ is the expected value.

The higher the correlation value, the more similar the two images are. Thus, NCC needs to be maximized.

SSD or NCC will probably not lead to satisfying results when comparing multimodal images like in Figure 7.9. Images from different scanning techniques can feature different structures, or corresponding structures do not necessarily show the same intensity. A suitable measure in such a case is mutual information.

Definition 7 (MI) *The* mutual information (MI) *of two images $\mathcal{T}$ and $\mathcal{R}$ is defined as*

$$\mathrm{MI}(\mathcal{T}, \mathcal{R}) := H(\mathcal{T}) + H(\mathcal{R}) - H(\mathcal{T}, \mathcal{R}) \, , \tag{7.18}$$

where, for an image I, $H(I) = -\sum p_I \log(p_I)$ is the entropy and p_I the relative histogram (or probability distribution) of I. $H(\mathcal{T}, \mathcal{R})$ is the entropy based on the joint relative histogram of $\mathcal{T}$ and $\mathcal{R}$.

The entropy measures the (joint) information of the input images $\mathcal{T}$ and $\mathcal{R}$. As MI measures the amount of shared information of two images it is usually a good choice for multimodality tasks. MI is high if the images share similar information and therefore has to be maximized.

Another criterion which is suitable for multimodality registration is the normalized gradient fields measure.

Definition 8 (NGF) *The* normalized gradient fields (NGF) *measure of two images $\mathcal{T}$ and $\mathcal{R}$ is defined as*

$$\mathrm{NGF}(\mathcal{T}, \mathcal{R}) := - \int_\Omega \langle n(\mathcal{T}, x), n(\mathcal{R}, x) \rangle^2 \, dx \, , \tag{7.19}$$

TABLE 7.1: Similarity measures and their field of application.

Measure	Monomodal	Multimodal
SSD	x	-
NCC	x	-
MI	x	x
NGF	x	x

where n is the normalized gradient

$$n(\mathcal{I}, i) := \begin{cases} \frac{\nabla \mathcal{I}(x)}{\sqrt{\|\nabla \mathcal{I}(x)\|^2 + \eta^2}} , & \nabla \mathcal{I}(x) \neq 0 \\ 0 & \text{otherwise} . \end{cases} \tag{7.20}$$

$\eta \in \mathbb{R}^{\geq 0}$ *is an edge parameter.*

For NGF it is assumed that the gradients of the template and reference image match when the images are perfectly aligned. Edges with a low gradient are treated as noise. This is controlled by the edge parameter η in Equation (7.20) which keeps the normalized gradient relatively low in such cases. This prevents low gradients in the images from dominating the registration results. As NGF does not operate directly on the image intensities, it is suitable for monomodal and multimodal registration (see [26] for details).

The four different similarity measures presented here are summarized and categorized according to their suitability for monomodality and multimodality in Table 7.1.

Other approaches

Besides the similarity measures discussed above, other approaches exist. An extension of MI called *conditional mutual information* (cMI) is proposed in [40], where a spatial component of the joint intensity pair is incorporated into MI. Another information-theoretic measure called *cross-cumulative residual entropy* (CCRE), which is a measure of entropy using cumulative distributions, was proposed in [67]. In contrast to fixed similarity measures, a learning similarity criterion, derived from max-margin structured output learning, was published in [35].

7.4.2 Validation

For evaluating newly developed methods, validation is essential. A comparison with existing methods as well as a quantitative and objective validation is required. Algorithms can be validated and quantified according to

1. Precision/Accuracy

2. Robustness

3. Speed/Complexity/Resource requirements

4. Clinical use

5. Assumption verification

In some cases the ground-truth information about the transformation is given, e.g., in phantom studies (see Section 7.7.2). Based on this information a magnitude and orientation error can be computed to quantify the accuracy.

Definition 9 (E^{MAG}) *For a transformation t the (averaged) magnitude error with regard to the ground-truth transformation t^{gt} is defined as*

$$E^{MAG}(t, t^{gt}) := \frac{1}{|\Omega|} \int_\Omega \left| \log\left(\frac{\|t(x)\|}{\|t^{gt}(x)\|}\right) \right| \, dx \,, \tag{7.21}$$

where $|\Omega|$ denotes $\int_\Omega 1 \, dx$.

The magnitude error measures the (average) error in length. The optimal value for E^{MAG} is 0. E^{MAG} is not bounded above.

Definition 10 (E^O) *For a transformation t the (averaged) orientation error with regard to the ground-truth transformation t^{gt} is defined as*

$$E^O(t, t^{gt}) := \frac{1}{2|\Omega|} \int_\Omega 1 - \frac{\langle t(x), t^{gt}(x)\rangle + 1}{\sqrt{(\|t(x)\|^2 + 1)\cdot(\|t^{gt}(x)\|^2 + 1)}} \, dx \,, \tag{7.22}$$

where $|\Omega|$ denotes $\int_\Omega 1 \, dx$.

The orientation error E^O indicates deviations regarding the angle and is scaled between 0 and 1. Again, the optimal value is 0.

An alternative definition of the orientation error in degrees is:

Definition 11 (E^O_{deg}) *For a transformation t the (averaged) orientation error in degrees with regard to the ground-truth transformation t^{gt} is defined as*

$$E^O_{deg}(t, t^{gt}) := \frac{1}{|\Omega|} \int_\Omega \arccos\left(\frac{\langle t(x), t^{gt}(x)\rangle + 1}{\sqrt{(\|t(x)\|^2 + 1)\cdot(\|t^{gt}(x)\|^2 + 1)}}\right) \, dx \,. \tag{7.23}$$

E^O_{deg} is scaled between 0° and 180° and the optimal value is 0°.

Since for real life data, usually no ground-truth information about the transformation is known, the precision and accuracy of the registration result has to be measured using image intensities. The lack of ground-truth information for real data asks for software and hardware phantoms with known transformations. See Section 7.7 and [70] for a more detailed discussion about validation.

7.4.3   Software

To develop new registration approaches it is often useful to build them on existing registration toolboxes. Technical and time-consuming tasks, such as implementing a multi-resolution framework, standard distance measures and loading, saving and plotting of images, can thus be avoided. Regarding standard registration tasks, it is advisable to use validated and reliable software tools. Therefore a small and incomplete list of freely (for non-commercial use) available registration software is given.

The ITK (Insight Toolkit, http://www.itk.org/) is a registration and segmentation framework written in C++. ITK is a suitable tool to implement new approaches or to use existing tools from the comprehensive collection. A registration toolbox based on ITK was recently published and is called *elastix* [34].

FAIR (Flexible Algorithms for Image Registration, http://www.cas. mcmaster.ca/~modersit/FAIR/) is a MATLAB® toolbox with an accompanying book [43]. Detailed examples are available with the software package. The basic distance measures and registration techniques already come along with the package. It is also a good framework to develop techniques.

The FSL (FMRIB Software Library, http://www.fmrib.ox.ac.uk/fsl/) software package is mainly developed for aligning MRI brain images. FSL contains the FLIRT tool for linear registration and the FNIRT tool (only SSD) for non-linear registration. The registration can be started from the command line or using a graphical user interface (GUI). FSL is made for direct use rather than for developing.

AIR (Automated Image Registration, http://bishopw.loni.ucla.edu/ AIR5/) can perform intra- and intermodal registration tasks. It has been validated for linear and non-linear PET-PET and PET-MRI registration.

ART (Automatic Registration Toolbox, http://www.nitrc.org/projects/ art) is made for linear and non-linear registration of brain images. Similar to FSL it is mainly used for aligning MRI images, but also for MRI-PET/SPECT registration.

In Table 7.2 the presented software packages are listed with some details about important features. These features are based on the following questions: Is the source code available? Which programming language is used? On which platform does the code run? Is a graphical user interface (GUI) available? Are examples included in the package?

7.5   Partial volume correction

Besides noise, as covered in Section 7.2, the partial volume effect (PVE) is a source of image degradation in medical imaging in general. The causes of

TABLE 7.2: Software

Software	Source	Language	Platform	GUI	Examples
ITK	yes	C++	Linux, PC, MAC	no	yes
FAIR	yes	MATLAB	Linux, PC, MAC	no	yes
FSL	yes	C++	Linux, PC, MAC	yes	yes
AIR	yes	C	Linux, PC, MAC	yes	no
ART	no	C++	Linux, PC, MAC	no	no

the PVE lie in the limited resolution of the respective imaging devices. Structures smaller than the system's resolution cannot be resolved which results in blurred boundaries for example. If the size of the structures to be examined is close to the imaging system's resolution, e.g., in imaging of small tumors, the impact of the PVE cannot be neglected.

This section deals with the mechanisms of the PVE and its impact on PET imaging (Section 7.5.1) as well as correction methods reducing the image degradation caused by PVE (Section 7.5.2).

7.5.1   The partial volume effect in PET imaging

Compared to morphological scanning techniques, one of the most decisive properties of PET is the quantification of tracer uptake *in vivo*. Tracer uptake can be quantified by means of the *standardized uptake value* (SUV), a normalized index taking different aspects into account, e.g., tracer dose, body weight and radioactive decay. The SUV allows a more precise diagnosis in tumor imaging and gives reliable information, e.g., about the progress of tumor therapies. The partial volume effect, however, makes it difficult to take full advantage of the potential of PET and has a strong impact on the accuracy of tracer concentration measurements like the SUV, particularly in small structures. The extent of this effect depends strongly on the resolution of the PET scanner.

In the following we show that partial volume correction is necessary to improve the quality of PET images and hence the quantification of tracer uptake. The influence of the PVE can be described by the image degradation model (see Equation (7.1))

$$I = P \star I_u + N \ . \tag{7.24}$$

Here, I is the PET image after reconstruction, I_u is the clean and uncorrupted image of the scanned object and N is additive noise. P denotes the *point spread function* which is, in general, the imaging system's response to a point source, i.e., how the system depicts an object smaller than the system's resolution. The signal acquired by scanning a point source resembles a Gaussian function (a Gaussian function is thus often used in practice to represent the PSF). The width of that signal at half its highest value is the *full width at half maximum* (FWHM), describing the system's resolution. Current human PET imaging

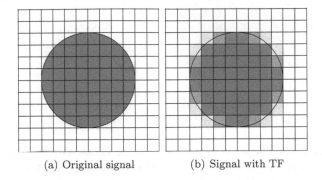

(a) Original signal (b) Signal with TF

FIGURE 7.11: (a) Signal with true borders *before* sampling. (b) Signal intensities *after* sampling at the voxel grid (with tissue fraction).

systems have a FWHM, i.e., spatial resolution of approximately 4–6 mm, small animal PET systems in preclinical research achieve a FWHM of approximately 1–2 mm.

The blurring induced by PVE is in fact caused by two different effects, which are often generalized as partial volume effect in the literature. These two effects, tissue fraction and spill-over, are described in the following paragraphs.

Tissue fraction

The causes of *tissue fraction* (TF) can be traced back to the process of sampling. The measured continuous signal is sampled at a finite voxel grid to create a three-dimensional PET image. Whenever the border of two adjacent tissues with different tracer concentrations lies inside a voxel of this grid, one can expect an erroneous contour. The intensity of such voxels is an average value of the different tracer concentrations inside the respective voxels.

Figure 7.11 shows a two-dimensional example to illustrate the impact of TF at the borderline of two neighboring tissues. Figure 7.11(b) shows the result of sampling the measured tracer concentrations indicated by the circle in Figure 7.11(a). The tissue is blurred and pixels at the tissue border show averaged intensity values, since they contain the signal of different compartments.

To summarize, the TF causes blurred edges but is significant only at tissue borders, since these voxels often contain strongly varying intensities. Thus, it is considered less severe compared to the *spill-over effect*.

Spill-over

The largest contribution to PVE is caused by the spill-over effect. As the name suggests, it can be described as a spilling of the measured tracer concentration of a voxel into its surrounding region. This leads to larger and dimmer structures in the PET image. One reason for this blurring is the finite

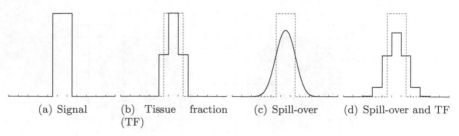

(a) Signal   (b) Tissue fraction   (c) Spill-over   (d) Spill-over and TF
                (TF)

FIGURE 7.12: Partial volume effects for a 1D signal. Tics on x-axis denote sampling intervals. (a) Input signal. (b) Due to tissue fraction the signal is an average of the intensities within the respective sampling area. (c) Spill-over causes part of the signal intensity to appear outside of the area of the true signal. (d) Combined effect of tissue fraction and spill-over. The input signal is plotted as a dashed line in (b)–(d) for orientation.

spatial resolution of imaging systems, e.g., the size and distance of detector crystals [51]. Besides physical limitations of PET scanners, the process of reconstruction contributes to the PVE as well.

The spill-over effect affects a voxel's intensity twofold: first, a voxel distributes part of its own signal to the surrounding region. Secondly, the voxel gets signal intensity from its neighbors. The following equation formalizes this relationship:

$$I(x) = I_u(x) - I_{out}(\Omega_x) + I_{in}(\Omega_x) \ . \tag{7.25}$$

The intensity of a voxel x in the reconstructed PET image I is composed of the actual intensity I_u, a spill-out I_{out} of intensity into the neighborhood Ω_x and a spill-in I_{in} from the neighboring voxels. Thus, the observed intensities are a mixture of the true intensities. Without knowledge of the exact amount of spill-in and spill-out it is not possible to reconstruct the true signal of a voxel. Many existing correction methods try to estimate the unknowns I_{in} and I_{out} in Equation (7.25) to compute the true signal.

Figure 7.12 shows the effects of tissue fraction and spill-over for a 1D signal (Figure 7.12(a)): Tissue fraction causes an averaging of all intensities within each sampling bin (Figure 7.12(b)). Spill-over causes part of the intensity to be visible outside of the area of the original signal (Figure 7.12(c)). The combination of both effects is illustrated in Figure 7.12(d).

7.5.2   Correction methods

In the following, methods used for *partial volume correction* (PVC) are discussed and we introduce a few representative algorithms. PVC is not yet commonly used in clinical environments, despite the great effort spent on PVC research in recent years. Yet, the importance of PVC for quantification has been shown [3, 65].

Recovery coefficient

The recovery coefficient (RC) is a multiplicative factor that is applied to a specified *region-of-interest* (ROI), e.g., a tumor or lesion, to correct the intensity values in this area. This factor has to be calculated for a structure that approximately equals the ROI in shape and size. The RC depends on the PET system and on the respective position within the field of view. It is calculated according to

$$ RC = \frac{\text{Measured sphere activity} - \text{Measured background activity}}{\text{Known sphere activity} - \text{Known background activity}} . \quad (7.26) $$

Computation of the RC is done by measuring the extent of the PSF for known signals of different sizes. This is compared to the ideal values of the ROIs. For spherical ROIs this leads to a function depending on the sphere size and the sphere-to-background contrast [63]. There are also ways to calculate the RC for non-spherical ROIs [62].

All RC algorithms are based on a compartment model with only two different tissue types, e.g., a tumor and the region near it. The RC algorithms can be divided into two groups: First, algorithms assuming the ROI is a *hot* spot in front of a *cold* background. Second, algorithms generalizing this assumption and allowing ROIs to be surrounded by regions with significant SUV values [63].

The use of RCs is rather limited for observing tumor metabolism following cancer therapy. This is due to the assumption of uniform uptake in the region-of-interest, which is not correct for tumors with necrotic tissue parts and can lead to biases in quantification [62].

Commonly, a tumor is assumed to be approximately spheric and its diameter can be well estimated from CT data. For this simple scenario clinical feasibility studies have been discussed recently [19].

Geometric transfer matrix

The *geometric transfer matrix* (GTM) is a generalization of the RC and expands the compartment model by using more than two different tissues. Yet, it is assuming homogeneous uptake for each region as well. The compartments have to be delineated either manually or automatically. Subsequently, each of the n compartments in the image is blurred by convolution with the system's PSF to simulate the spill-over effect to the remaining $n - 1$ compartments. This yields a $n \times n$ matrix, called the geometric transfer matrix, composed of transfer coefficients W_{ij}. The coefficients W_{ij} give the signal spill-over from compartment i to compartment j. The PVC problem can be restated as a linear equation system of n unknowns and n equations:

$$ W \cdot u = m , \quad (7.27) $$

where $m \in \mathbb{R}_+^n$ is the vector of signal intensities for the n compartments and where $u \in \mathbb{R}_+^n$ denotes the unknown vector of partial volume corrected compartments. The solution of this system gives an estimation of the true signal for each compartment without the influence of the spill-over effect.

A disadvantage of this approach lies in the need for an exact delineation of the individual compartments. In PET imaging the selection of functional regions is usually made using anatomical data taken from registered CT or MRI data [71]. Similar to the case of the RC, the uniform tracer uptake is a limiting assumption.

Deconvolution

Deconvolution tries to reverse the convolution of a clean image I_u with the PSF P (see Equation (7.24)). In the absence of additive noise N and given a known P, $I = P \star I_u$ can be deconvolved using the convolution theorem (Section 7.2.2). In real PET systems, however, noise is always present and the PSF may not be known exactly. This complicates the deconvolution process enormously. During direct deconvolution, noise is amplified strongly. We clarify this relationship with the help of the model of image degradation: Fourier transform of Equation (7.24) and application of the convolution theorem leads to

$$\mathcal{F}(I_u) = \frac{\mathcal{F}(I)}{\mathcal{F}(P)} - \frac{\mathcal{F}(N)}{\mathcal{F}(P)}, \tag{7.28}$$

where $\mathcal{F}$ denotes the Fourier transform. Noise is often located in the higher frequencies. Since P is a Gaussian function, the values of $\mathcal{F}(P)$ are close to zero at high frequencies. For this reason the noise $\mathcal{F}(N)$ gets amplified due to the division by $\mathcal{F}(P)$.

Figure 7.13 shows the results of different deconvolution techniques for an artificial example in the absence of noise. Iterative deconvolution algorithms try to avoid the amplification of additive noise, which is hard to put into practice. In the following we describe suitable approaches for partial volume correction based on deconvolution techniques.

Van-Cittert deconvolution

The van-Cittert iteration can be used to perform PVC if the point spread function is known approximately. The iteration scheme is given by

$$I_0 = I \tag{7.29}$$

$$I_{j+1} = I_j + \alpha(I_0 - P \star I_j), \tag{7.30}$$

where I is the given PET image and $\alpha \in \mathbb{R}^+$ a relaxation parameter. The criterion for ending the iteration scheme is crucial since the amount of noise increases with every iteration step [64].

Richardson–Lucy deconvolution

The Richardson–Lucy (RL) deconvolution uses a statistical approach. The PSF is assumed to be known approximately. The basic idea of the RL deconvolution is to correct the observed image I towards a maximum likelihood

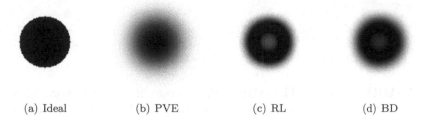

| (a) Ideal | (b) PVE | (c) RL | (d) BD |

FIGURE 7.13: Results of different deconvolution techniques on a 2D artificial example *without* noise. (a) A high-contrast sphere on uniform background. (b) Impact of the PVE, simulated by convolution with a Gaussian kernel. (c) and (d) PVC results of deconvolution using Richardson–Lucy and blind deconvolution, respectively.

solution. The iteration scheme is given by

$$I_0 = C, \text{ with } C > 0 \text{ a constant, uniform image} \qquad (7.31)$$

$$I_{j+1} = I_j \cdot \left(P^T \star \left(\frac{I_u}{P \star I_j} \right) \right) . \qquad (7.32)$$

Division and multiplication operations in this iteration scheme are component-wise. Further technical details can be found in [10]. The RL algorithm is a special case of *expectation maximization algorithms*. It is quite popular since no assumptions regarding amount or type of noise have to be made. Moreover, no regularization parameter has to be estimated.

A qualitative comparison of van-Cittert deconvolution and Richardson–Lucy algorithm for PET imaging revealed that the latter performs better with regard to noise amplification [6].

Blind deconvolution
For an unknown point spread function the previously described methods are not applicable. Approaches to estimate the PSF are known as blind deconvolution (BD) methods. The PSF can be estimated either iteratively or in a single step by using additional knowledge, e.g., anatomical information. In the next step the estimated PSF is used to perform the deconvolution using approaches like the Richardson–Lucy algorithm or the Wiener filter [21].

Since the mathematical theory of BD techniques is complex, no detailed discussion is given here. A BD algorithm for brain-SPECT is discussed in [42].

Figure 7.14 shows the results of different deconvolution techniques for real PET data. The original image in Figure 7.14(a) shows a liver tumor (dark structure). In contrast to the artificial example in Figure 7.13 the results of deconvolution suffer from noise.

    (a) Original   (b) RL   (c) BD

FIGURE 7.14: (a) The original PET image of a liver tumor. (b) and (c) Deconvolution results using Richardson–Lucy and blind deconvolution respectively.

Multiresolution

The inclusion of information from high-resolution images, e.g., CT or MRI, is a promising approach for PVC. The information of functional imaging, i.e., of PET data, is combined with the anatomical information of high resolution CT data in a reasonable way. Besides a registration of PET data and the respective modality, a positive correlation of intensity values in the region-of-interest is required. The additionally used imaging modality is downsampled to the resolution of PET, for example, by using wavelets. Thus, it is possible to reconstruct lost details in the low resolution image using the additional resolution levels from the wavelet transform. A comparison of the registered images results in partial volume corrected signal intensities.

    In [5] a multiresolution technique is proposed using 2D wavelet transforms and a linear model for scaling the transformed images. Based on this approach a method using Markov-modeling is proposed in [50] to compare the two image modalities in order to avoid the introduction of PVC artifacts. An improvement compared to [5] is reported in [50]. The advantage of these methods is that no prior knowledge about the PSF is required.

7.6   Super-resolution

    Compared to CT with a maximal spatial resolution below $100\,\mu$m, PET and SPECT obtain a relatively poor resolution of approximately $4\,$mm. Thus, there is a demand to increase resolution of PET. Besides improving the PET scanner's hardware, resolution is to a certain extent improvable through image processing methods. These methods estimate a high resolution image (HR) from several low resolution images (LR). The LR images are acquired at slightly shifted positions, thus supplying complementary image information. In the reconstruction of PET images the resolution obtained is usually lower than the scanner's resolution due to a trade-off between noise and resolution [25].

The applicability of super-resolution to PET has been shown, e.g., in [32] using an approach proposed in [30]. This approach is outlined briefly in the following to illustrate the general formalism.

Given K shifted LR images L_k, $k \in \{1, \ldots, K\}$, $K \in \mathbb{N}$, an initial guess of the HR image $H^{(0)}$ is obtained by averaging the upsampled LR images:

$$H^{(0)} = \frac{1}{K} \sum_{k=1}^{K} t_k^{-1}(L_k \uparrow s), \tag{7.33}$$

where t_k is the transformation of the shifted LR images to a common reference frame and $\uparrow s$ is the upsampling of the LR image to the resolution of the HR image. Based on this initial guess, low resolution images $\tilde{L}_k^{(n)}$ are estimated by means of

$$\tilde{L}_k^{(n)} = (t_k(H^{(n)}) \star h) \downarrow s, \tag{7.34}$$

where $\star$ is the convolution operator, $\downarrow s$ denotes downsampling to the resolution of the LR images, h is a smoothing kernel and (n) denotes the iteration. The difference between the estimated $\tilde{L}_k^{(n)}$ and the measured LR images L_k is used to update the estimate of the HR image:

$$H^{(n+1)} = H^{(n)} + \frac{1}{K} \sum_{k=1}^{K} t_k^{-1}(((L_k - \tilde{L}_k^{(n)}) \uparrow s) \star p), \tag{7.35}$$

where p is a sharpening kernel. Ideally, p is the inverse of the blurring kernel (if the inverse exists), or the approximate inverse. For details we refer to [30].

The procedure is repeated until either a fixed number of iterations is reached or the difference between the estimated and measured LR images is lower than a chosen threshold. In conclusion, the outlined method determines the HR image by estimating an image which produces the measured LR images L_k given the transformations t_k and the blurring kernel h.

Figure 7.15 gives an artificial example using the outlined method. Figure 7.15(b) shows one of the eight LR images generated from the original image in Figure 7.15(a). The smaller gaps in the lines in the LR image (Figure 7.15(b)) are not clearly resolvable. In the computed HR image (Figure 7.15(c)) the gaps in the lines are recognizable again.

In [66] improvements regarding SNR (see Definition 14 and 15) for respiratory gated phantom data using super-resolution are shown. A concise survey of super-resolution in medical imaging focusing on PET and MRI is given in [25].

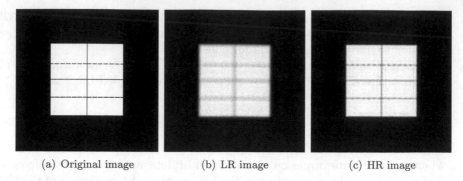

(a) Original image   (b) LR image   (c) HR image

FIGURE 7.15: Artificial super-resolution example. (a) Input image. (b) The LR images (only one shown) generated from the original image. (c) The HR image computed from the LR images.

7.7   Validation

When new image processing methods for emission tomography are developed one question automatically arises: How good is the method? The word "good" is ambiguous—it can mean the general accuracy, the accuracy compared to other methods, the computational complexity, or the clinical usefulness. Validation can be understood as an attempt to answer the above question.

Validation is essential in many fields of emission tomography. Registration methods (see Section 7.4), denoising techniques (see Section 7.2), partial volume correction (see Section 7.5), and reconstruction methods need to be validated, to name just a few.

In general, every application has its own case-specific validation methods. However, two universal ways of validation will be discussed in the following. Firstly, error measures are introduced in Section 7.7.1, as validation can often be understood as a pixel or voxel-based comparison of images. Secondly, phantoms providing ground-truth information are discussed in Section 7.7.2.

Another aspect of validation is a direct performance comparison with other popular methods. When bringing a method into clinical practice this is a mandatory step. Methods for training and testing, and useful statistical tests are discussed in detail in [7] and are not part of this section.

7.7.1   Intensity-based measures

A common situation in medical image processing is that images need to be judged or compared based on their intensity values. The similarity measures

SSD, NCC, MI, and NGF discussed in Section 7.4.1.2 are suitable for general intensity-based validation (see Definitions 5–8). In Section 7.4.2 some additional validation criteria are introduced with E^{MAG}, E^O, and E^O_{deg} to measure deviations in magnitude and orientation of vector fields.

To supplement the intensity-based criteria the relative error is defined. It measures the distance of two images $\mathcal{I}$ and $\mathcal{R}$ relative to the norm of the reference image $\mathcal{R}$.

Definition 12 (RE) *For an image $\mathcal{I}$ and a reference image $\mathcal{R}$ the relative error is defined as*

$$\text{RE}(\mathcal{I}, \mathcal{R}) := \int_\Omega \frac{(\mathcal{I}(x) - \mathcal{R}(x))^2}{\mathcal{R}(x)^2} \, dx \ . \tag{7.36}$$

The relative error (RE) is closely related to the SSD measure. The difference of RE compared to SSD is that it is normalized by the intensity of $\mathcal{R}$. Hence, differently scaled images have the same relative error: $\text{RE}(\mathcal{I}, \mathcal{R}) = \text{RE}(\alpha \cdot \mathcal{I}, \alpha \cdot \mathcal{R})$, $\alpha \in \mathbb{R}^+$. The relative error is minimal for $\text{RE} = 0$.

In some cases it is interesting to compare the relative similarity of two images before and after processing. This is expressed by the following alternative definition of the relative error:

Definition 13 (RE_{alt}) *For an image $\mathcal{I}$, a reference image $\mathcal{R}$ and the original version $\mathcal{I}_o$ of $\mathcal{I}$ before processing the alternative relative error is defined as (if $\mathcal{I}_o$ and $\mathcal{R}$ are not equal)*

$$\text{RE}_{alt}(\mathcal{I}, \mathcal{I}_o, \mathcal{R}) := \int_\Omega \frac{(\mathcal{I}(x) - \mathcal{R}(x))^2}{(\mathcal{I}_o(x) - \mathcal{R}(x))^2} \, dx \ . \tag{7.37}$$

If the similarity of $\mathcal{I}$ and $\mathcal{R}$ is comparable to the similarity of $\mathcal{I}_o$ and $\mathcal{R}$ (before processing), RE_{alt} is about 1. The higher the improvement of the similarity the closer RE_{alt} gets to 0.

In Section 7.2 it is stated that noise is omnipresent in medical imaging. In some cases it is necessary to quantify the noise level of an image. A noise level comparison before and after processing (e.g., filtering) is also often useful. For quantifying the noise level the signal to noise ratio (SNR) is a possible choice. As the noiseless image is generally unknown the SNR can be estimated according to the following definition:

Definition 14 (SNR) *For a non-constant image $\mathcal{I}$ the signal to noise ratio (SNR) can be defined as*

$$\text{SNR}(\mathcal{I}) := \frac{\mu(\mathcal{I})}{\sigma(\mathcal{I})} = \frac{\frac{1}{|\Omega|} \int_\Omega \mathcal{I}(x) dx}{\sqrt{\frac{1}{|\Omega|} \int_\Omega (\mathcal{I}(x) - \mu(\mathcal{I}))^2 dx}} \ , \tag{7.38}$$

where μ is the expected value and σ the standard deviation.

The SNR values for the images in Figure 7.1 from left to right are 0.8378 (Original), 0.8951 (Mean), 0.8514 (Median), and 0.8536 (Wiener), respectively. These example values show that all filtered images have a higher SNR than the original image, indicating a reduced noise level. Obviously, the lower the variance of the image is the higher the SNR gets.

The peak SNR is a further common variant of the SNR. It measures the relation between the maximum intensity of an image and its noise component.

Definition 15 (PSNR) *Let L define the maximum intensity of a nonconstant image $\mathcal{I}$. The peak signal to noise ratio (PSNR) is defined as*

$$\text{PSNR}(\mathcal{I}) := 20 \cdot \log_{10} \frac{L}{\sigma(\mathcal{I})} = 20 \cdot \log_{10} \frac{L}{\sqrt{\frac{1}{|\Omega|} \int_{\Omega} (\mathcal{I}(x) - \mu(\mathcal{I}))^2 dx}} , \quad (7.39)$$

where μ is the expected value and σ the standard deviation.

SNR and PSNR have in common that the denominator is the standard deviation. They differ in the numerator which is the expected value for SNR and the maximum intensity for PSNR. The logarithmic scaling in the definition of PSNR is used to express the noise ratio in decibels.

7.7.2 Phantoms

The validation of many image processing techniques is done using phantoms as they provide complete knowledge about *anatomy* and/or *physiology* and therefore ground-truth information. This is an advantage compared to real data which generally lacks ground-truth information.

Phantoms can be subdivided into *hardware* and *software* phantoms. Hardware phantoms are commonly used to validate scanner geometry or reconstruction software. Software phantoms, however, can be used for validation of image processing tasks. These two kinds of phantoms are discussed in the following.

7.7.2.1 Hardware

Hardware phantoms can be used for validating scanner geometry, attenuation and scatter correction or alignment tasks in multimodal imaging. Further, hardware phantoms are the established way to determine the scanner resolution. They are inevitable for quantitative measurements.

Most hardware phantoms available for emission tomography are commercial like Jaszczak or NEMA [2, 8, 14] phantoms. The commercially available hardware phantoms cover simple geometry phantoms or more detailed brain, cardiac, or lung phantoms. Phantoms applicable to multimodal imaging techniques like SPECT/CT or PET/CT are also available. Especially for combined approaches, where the coordinate systems of the different scanners need to be calibrated, phantoms are essential.

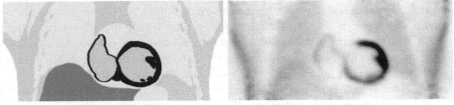

(a) Input phantom   (b) Simulated PET acquisition

FIGURE 7.16: Simulated PET acquisition using GATE. (a) Static XCAT phantom used as input for simulation (1 mm^3 voxel size). (b) Result from simulated PET acquisition using GATE. Simulated scan duration is 5 minutes.

7.7.2.2   Software

Software phantoms are helpful to develop and validate medical image processing algorithms as they provide morphological (anatomy) and/or functional (physiology) ground-truth information. In connection with software phantoms the realistic simulation of scanner acquisition processes is discussed. Popular phantoms in emission tomography are XCAT, NCAT, and MOBY which are often combined with the simulation tool GATE.

XCAT, NCAT, and MOBY
The XCAT, NCAT, and MOBY phantoms [56] allow the generation of PET images and attenuation maps. While XCAT and NCAT are based on human data, MOBY is a mouse phantom. All phantoms provide a model of the subject's anatomy and physiology. It is possible to include respiratory and cardiac motion resulting in 4D images. To simulate tracer dynamics the additional use of time activity curves (TACs) is possible. The outcome is idealized noiseless data.

GATE
GATE (Geant4 Application for Tomographic Emission) [31] is a Monte Carlo simulation-based toolkit for generating simulated PET or SPECT data under realistic acquisition conditions. For example source or detector movement and source decay kinetics can be modeled. GATE is a pure simulation framework and does not include any predefined patient models. The output of XCAT, NCAT, or MOBY can be used as input for GATE to simulate a realistic scanner acquisition. Figure 7.16 gives an example of a simulated PET acquisition using GATE and the XCAT phantom.

Related work
The Zubal phantom [73] is a human head and torso phantom which is commonly used for many modeling and simulation calculations. It is a common alternative to the XCAT and NCAT phantom.

A labeled whole-body atlas for mice is the Digimouse dataset [16]. The package covers the registered voxelized source data as well as surface and volume tessellations. In [28] a method for reconstructing 3D atlases from digital 2D atlas diagrams is presented. Tracer-specific templates were created in [11] using small-animal PET data mapped onto a 3D MRI rat brain template which is oriented according to the rat brain Paxinos atlas.

Apart from quantitative algorithm validation there is a great demand for atlas and phantom based methods, e.g., in atlas-based attenuation correction [29, 44] and PVC [58].

In conclusion, to reliably validate new image processing methods it is essential to take advantage of hardware or software phantoms. The ground-truth information provided by phantoms is often compared to the methods' results using intensity based measures.

References

[1] E.D. Angelini, J. Kalifa, and A.F. Laine. Harmonic multiresolution estimators for denoising and regularization of SPECT-PET data. In *IEEE International Symposium on Biomedical Imaging*, pages 697–700, 2002.

[2] National Electrical Manufacturers Association. NEMA standards publication NU 2-2001: Performance measurements of positron emission tomographs. National Electrical Manufacturers Association, 2001.

[3] N. Avril, S. Bense, S.I. Ziegler, J. Dose, W. Weber, C. Laubenbacher, W. Rmer, F. Jnicke, and M. Schwaiger. Breast imaging with Fluorine-18-FDG PET: Quantitative image analysis. *Journal of Nuclear Medicine*, 38(8):1186–1191, August 1997.

[4] I.N. Bankman, editor. *Handbook of Medical Image Processing and Analysis*. Academic Press, Inc., Burlington, MA, USA, 2nd edition, 2009.

[5] N. Boussion, M. Hatt, F. Lamare, Y. Bizais, A. Turzo, C. Cheze-Le Rest, and D. Visvikis. A multiresolution image based approach for correction of partial volume effects in emission tomography. *Physics in Medicine and Biology*, 51(7):1857–1876, April 2006.

[6] N. Boussion, C. Cheze-Le Rest, M. Hatt, and D. Visvikis. Incorporation of wavelet-based denoising in iterative deconvolution for partial volume correction in whole-body PET imaging. *European Journal of Nuclear Medicine and Molecular Imaging*, 36(7):1064–75, 2009.

[7] K.W. Bowyer. Validation of medical image analysis techniques. In *Handbook of Medical Imaging, Volume 2. Medical Image Processing and Analysis*, 1st edition. pages 567–608. SPIE, June 2000.

[8] G. Brix, J. Zaers, L.E. Adam, M.E. Bellemann, H. Ostertag, H. Trojan, U. Haberkorn, J. Doll, F. Oberdorfer, and W. Lorenz. Performance evaluation of a whole-body PET scanner using the NEMA protocol. *Journal of Nuclear Medicine*, 38(10):1614–1623, 1997.

[9] L.G. Brown. A survey of image registration techniques. *ACM Computing Surveys*, 24(4):325–376, 1992.

[10] A. S. Carasso. Linear and nonlinear image deblurring: A documented study. *SIAM Journal on Numerical Analysis*, 36:1659–1689, 1999.

[11] C. Casteels, P. Vermaelen, J. Nuyts, A. Van Der Linden, V. Baekelandt, L. Mortelmans, G. Bormans, and K. Van Laere. Construction and evaluation of multitracer small-animal PET probabilistic atlases for voxel-based functional mapping of the rat brain. *Journal of Nuclear Medicine*, 47(11):1858–1866, 2006.

[12] F. Sanchez Castro, C. Pollo, J. Villemure, and J. Thiran. Feature-segmentation-based registration for fast and accurate deep brain stimulation targeting. In *Proceedings of the 20th International Congress and Exhibition in Computer Assisted Radiology and Surgery*, Parallel Computing in Electrical Engineering. IEEE, 2006.

[13] W.R. Crum, T. Hartkens, and D.L.G. Hill. Non-rigid image registration: theory and practice. *British Journal of Radiology*, 77(2):S140–153, 2004.

[14] M.E. Daube-Witherspoon, J.S. Karp, M.E. Casey, F.P. DiFilippo, H. Hines, G. Muehllehner, V. Simcic, C.W. Stearns, L.E. Adam, S. Kohlmyer, and V. Sossi. PET performance measurements using the NEMA NU 2-2001 standard. *Journal of Nuclear Medicine*, 43(10):1398–1409, 2002.

[15] M. Dawood, C. Brune, X. Jiang, F. Büther, M. Burger, O. Schober, M. Schäfers, and K.P. Schäfers. A continuity equation based optical flow method for cardiac motion correction in 3D PET data. In *Proceedings of 5th International Conference on Medical Imaging and Augmented Reality*, *LNCS*, volume 6326, pages 88–97, 2010.

[16] B. Dogdas, D. Stout, A.F. Chatziioannou, and R.M. Leahy. Digimouse: a 3D whole body mouse atlas from CT and cryosection data. *Physics in Medicine and Biology*, 52(3):577–587, 2007.

[17] B. Fischer and J. Modersitzki. Combination of automatic non-rigid and landmark based registration: the best of both worlds. In M. Sonka and J.M. Fitzpatrick, editors, *Medical Imaging 2003: Image Processing*, volume 5032, pages 1037–1048. SPIE, 2003.

[18] B. Fischer and J. Modersitzki. Ill-posed medicine—an introduction to image registration. *Inverse Problems*, 24(3):1–19, 2008.

[19] L. Geworski, B. O. Knoop, M. Levi de Cabrejas, W. H. Knapp, and D. L. Munz. Recovery correction for quantitation in emission tomography: a feasibility study. *European Journal of Nuclear Medicine*, 27(2):161–169, February 2000.

[20] F. Gigengack, L. Ruthotto, M. Burger, C.H. Wolters, X. Jiang, and K.P. Schaefers. Motion correction of cardiac PET using mass-preserving registration. In *NSS/MIC Conference Record, IEEE*, 2010.

[21] R.C. Gonzalez and R.E. Woods. *Digital Image Processing*. Prentice-Hall, Inc., Upper Saddle River, NJ, USA, 3rd edition, 2006.

[22] R.C. Gonzalez, R.E. Woods, and S.L. Eddins. *Digital Image Processing Using MATLAB*. Prentice-Hall, Inc., Upper Saddle River, NJ, USA, 2003.

[23] A.A. Goshtasby. *2-D and 3-D Image Registration: For Medical, Remote Sensing, and Industrial Applications*. Wiley-Interscience, 2005.

[24] G. Green. Wavelet-based denoising of cardiac PET data. Master's thesis, Department of Systems and Computer Engineering, Carleton University, 2005.

[25] H. Greenspan. Super-resolution in medical imaging. *The Computer Journal*, 52(1):43–63, February 2009.

[26] E. Haber and J. Modersitzki. Intensity gradient based registration and fusion of multi-modal images. *Medical Image Computing and Computer-Assisted Intervention (MICCAI)*, pages 726–733, 2006.

[27] J.V. Hajnal, D.L.G. Hill, and D.J. Hawkes. *Medical Image Registration*. CRC Press, Boca Raton, FL, 2001.

[28] T. Hjornevik, T.B. Leergaard, D. Darine, O. Moldestad, A.M. Dale, F. Willoch, and J.G. Bjaalie. Three-dimensional atlas system for mouse and rat brain imaging data. *Frontiers in Neuroinformatics*, 1:1–11, 2007.

[29] M. Hofmann, F. Steinke, V. Scheel, G. Charpiat, J. Farquhar, P. Aschoff, M. Brady, B. Schölkopf, and B.J. Pichler. MRI-based attenuation correction for PET/MRI: A novel approach combining pattern recognition and atlas registration. *Journal of Nuclear Medicine*, 49(11):1875–1883, 10 2008.

[30] M. Irani. Motion analysis for image enhancement: Resolution, occlusion, and transparency. *Journal of Visual Communication and Image Representation*, 4(4):324–335, December 1993.

[31] S. Jan, G. Santin, D. Strul, S. Staelens, K. Assi, D. Autret, S. Avner, R. Barbier, M. Bardis, P.M. Bloomfield, D. Brasse, V. Breton, P. Bruyndonckx, I. Buvat, A.F. Chatziioannou, Y. Choi, Y.H. Chung, C. Comtat,

D. Donnarieix, L. Ferrer, S.J. Glick, C.J. Groiselle, D. Guez, P.-F. Honore, S. Kerhoas-Cavata, A.S. Kirov, V. Kohli, M. Koole, M. Krieguer, D.J. van der Laan, F. Lamare, G. Largeron, C. Lartizien, D. Lazaro, M.C. Maas, L. Maigne, F. Mayet, F. Melot, C. Merheb, E. Pennacchio, J. Perez, U. Pietrzyk, F.R. Rannou, M. Rey, D.R. Schaart, C.R. Schmidtlein, L. Simon, T.Y. Song, J.-M. Vieira, D. Visvikis, R. Van de Walle, E. Wiers, and C. Morel. GATE: A simulation toolkit for PET and SPECT. *Physics in Medicine and Biology*, 49(19):4543–4561, 2004.

[32] J.A. Kennedy, O. Isracl, A. Frenkel, R. Bar-Shalom, and H. Azhari. Super-resolution in PET imaging. *IEEE Transactions on Medical Imaging*, 25(2):137–147, February 2006.

[33] A. Klein, J. Andersson, B.A. Ardekani, J. Ashburner, B. Avants, M.-C. Chiang, G.E. Christensen, D.L. Collins, J. Gee, P. Hellier, J.H. Song, M. Jenkinson, C. Lepage, D. Rueckert, P. Thomson, T. Vercauteren, R.P. Woods, J.J. Mann, and R. Parsey. Evaluation of 14 nonlinear deformation algorithms applied to human brain MRI registration. *NeuroImage*, 46(3):786–802, 2009.

[34] S. Klein, M. Staring, K. Murphy, M.A. Viergever, and J.P.W. Pluim. elastix: a toolbox for intensity-based medical image registration. *IEEE Transactions on Medical Imaging*, 29(1):196–205, January 2010.

[35] D. Lee, M. Hofmann, F. Steinke, Y. Altun, N.D. Cahill, and B. Schölkopf. Learning similarity measure for multi-modal 3D image registration. *IEEE Computer Society Conference on Computer Vision and Pattern Recognition*, pages 186–193, 2009.

[36] T.M. Lehmann, C. Gönner, and K. Spitzer. Survey: Interpolation methods in medical image processing. *IEEE Transactions on Medical Imaging*, 18:1049–1075, 1999.

[37] H. Lester and S.R. Arridge. A survey of hierarchical non-linear medical image registration. *Pattern Recognition*, 32(1):129–149, January 1999.

[38] G. Li, H. Xie, H. Ning, J. Capala, B.C. Arora, C.N. Coleman, K. Camphausen, and R.W. Miller. A novel 3D volumetric voxel registration technique for volume-view-guided image registration of multiple imaging modalities. *International Journal of Radiation Oncology · Biology · Physics*, 63(1):261–273, 2005.

[39] J.S. Lim. *Two-dimensional Signal and Image Processing*. Prentice-Hall, Inc., Upper Saddle River, NJ, USA, 1990.

[40] D. Loeckx, P. Slagmolen, F. Maes, D. Vandermeulen, and P. Suetens. Nonrigid image registration using conditional mutual information. *IEEE Transactions on Medical Imaging*, 29(1):19–29, May 2009.

[41] J.B.A. Maintz and M.A. Viergever. A survey of medical image registration. *Medical Image Analysis*, 2(1):1–36, 1998.

[42] M. Mignotte and J. Meunier. Three-dimensional blind deconvolution of SPECT images. *IEEE Transactions on Biomedical Engineering*, 47(2):274–280, January 2000.

[43] J. Modersitzki. *FAIR: Flexible Algorithms for Image Registration (Fundamentals of Algorithms)*. Society for Industrial and Applied Mathematics, October 2009.

[44] M. Montandon and H. Zaidi. Atlas-guided non-uniform attenuation correction in cerebral 3D PET imaging. *NeuroImage*, 25(1):278–86, 2005.

[45] M.C. Motwani, M.C. Gadiya, R.C. Motwani, and Jr. F.C. Harris. Survey of image denoising techniques. *Proceedings of GSPx*, 2004.

[46] A. Neumaier. Solving ill-conditioned and singular linear systems: A tutorial on regularization. *SIAM Review*, 40(3):636–666, 1998.

[47] J. Nocedal and S.J. Wright. *Numerical Optimization (Springer Series in Operations Research and Financial Engineering)*. Springer, 2nd edition, July 2006.

[48] M. Petrou and P. Bosdogianni. *Image Processing: The Fundamentals*. John Wiley & Sons, Ltd, October 2001.

[49] J.P.W. Pluim, J.B.A. Maintz, and M.A. Viergever. Interpolation artefacts in mutual information-based image registration. *Computer Vision and Image Understanding*, 77(9):211–232, 2000.

[50] A. Le Pogam, M. Hatt, N. Boussion, D. Guilloteau, J. L. Baulieu, C. Prunier, F. Turkheimer, and D. Visvikis. Conditional partial volume correction for emission tomography: A wavelet-based hidden Markov model and multi-resolution approach. In *5th IEEE International Symposium on Biomedical Imaging: From Nano to Macro*, pages 1319–1322, 2008.

[51] O. Rousset, A. Rahmim, A. Alavi, and H. Zaidi. Partial volume correction strategies in PET. *PET Clinics*, 2(2):235–249, April 2007.

[52] J.C. Russ. *The Image Processing Handbook*. CRC Press, Inc., Boca Raton, FL, USA, 5th edition, 2007.

[53] O. Salvado, C.M. Hillenbrand, and D.L. Wilson. Partial volume reduction by interpolation with reverse diffusion. *International Journal of Biomedical Imaging*, pages 1–13, 2006.

[54] M. Scharfe, R. Pielot, and F. Schreiber. Fast multi-core based multimodal registration of 2D cross-sections and 3D datasets. *BMC Bioinformatics*, 11(1):20, 2010.

[55] O. Scherzer. *Mathematical Models for Registration and Applications to Medical Imaging.* Springer, New York, 2006.

[56] W.P. Segars and B.M.W. Tsui. The MCAT, NCAT, XCAT, and MOBY Computational Human and Mouse Phantoms. In *Handbook of Anatomical Models for Radiation Dosimetry*, eds. X.G. Xu and K.F. Eckerman, 105–134, Boca Raton, FL: Taylor & Francis, 2010.

[57] R. Shekhar, V. Walimbe, S. Raja, V. Zagrodsky, M. Kanvinde, G. Wu, and B. Bybel. Automated 3-dimensional elastic registration of whole-body PET and CT from separate or combined scanners. *Journal of Nuclear Medicine*, 46(9):1488–1496, 2005.

[58] M. Shidahara, C. Tsoumpas, A. Hammers, N. Boussion, D. Visvikis, T. Suhara, I. Kanno, and F.E. Turkheimer. Functional and structural synergy for resolution recovery and partial volume correction in brain PET. *NeuroImage*, 44(2):340–348, 2009.

[59] Y.Y. Shih, J.C. Chen, and R.S. Liu. Development of wavelet de-noising technique for PET images. *Computerized Medical Imaging and Graphics*, 29(4):297–304, 2005.

[60] P.J. Slomka, D. Dey, C. Przetak, U.E. Aladl, and R.P. Baum. Automated 3-dimensional registration of stand-alone ^{18}F-FDG whole-body PET with CT. *Journal of Nuclear Medicine*, 44(7):1156–1167, 2003.

[61] M. Sonka and J.M. Fitzpatrick. *Handbook of Medical Imaging, Volume 2. Medical Image Processing and Analysis.* SPIE, 1st edition, June 2000.

[62] M. Soret, S.L. Bacharach, and I. Buvat. Partial-volume effect in PET tumor imaging. *Journal of Nuclear Medicine*, 48(6):932–945, June 2007.

[63] S. M. Srinivas, T. Dhurairaj, S. Basu, G. Bural, S. Surti, and A. Alavi. A recovery coefficient method for partial volume correction of PET images. *Annals of Nuclear Medicine*, 23(4):341–348, June 2009.

[64] B. K. Teo, Y. Seo, S. L. Bacharach, J. A. Carrasquillo, S. K. Libutti, H. Shukla, B. H. Hasegawa, R. A. Hawkins, and B. L. Franc. Partial-volume correction in PET: Validation of an iterative postreconstruction method with phantom and patient data. *Journal of Nuclear Medicine*, 48(5):802–810, May 2007.

[65] P. Tylski, S. Stute, N. Grotus, K. Doyeux, S. Hapdey, I. Gardin, B. Vanderlinden, and I. Buvat. Comparative assessment of methods for estimating tumor volume and standardized uptake value in 18F-FDG PET. *Journal of Nuclear Medicine*, 51(2):268–276, February 2010.

[66] D. Wallach, F. Lamare, J. Rubio, M. J. Ledesma-Carbayo, G. Kontaxakis, A. Santos, P. Marechal, C. Roux, and D. Visvikis. Super-resolution in

4d positron emission tomography. In *IEEE Nuclear Science Symposium Conference Record*, pages 4285–4287. IEEE, October 2008.

[67] F. Wang and B.C. Vemuri. Non-rigid multi-modal image registration using cross-cumulative residual entropy. *International Journal of Computer Vision*, 74(2):201–215, 2007.

[68] J. Weickert and M. Welk. *Visualization and Processing of Tensor Fields*, chapter Tensor Field Interpolation with PDEs, pages 315–325. Springer, 2006.

[69] T.Z. Wong, T.G. Turkington, T.C. Hawk, and R.E. Coleman. PET and brain tumor image fusion. *Cancer Journal*, 10(4):234–42, 2004.

[70] R.P. Woods. *Handbook of Medical Image Processing and Analysis*, chapter Validation of Registration Accuracy, pages 569–575. Academic Press, Inc., Burlington, MA, USA, 2nd edition, 2009.

[71] H. Zaidi, T. Ruest, F. Schoenahl, and M. L. Montandon. Comparative assessment of statistical brain MR image segmentation algorithms and their impact on partial volume correction in PET. *NeuroImage*, 32:161–169, October 2006.

[72] B. Zitova and J. Flusser. Image registration methods: a survey. *Image and Vision Computing*, 21(11):977–1000, 2003.

[73] I.G. Zubal, C.R. Harrell, E.O. Smith, Z. Rattner, G. Gindi, and P.B. Hoffer. Computerized three-dimensional segmented human anatomy. *Medical Physics*, 21(2):299–302, 1994.

Chapter 8

Motion Correction in Emission Tomography

Mohammad Dawood
European Institute for Molecular Imaging, University of Münster, Münster, Germany

8.1   Introduction 157
    8.1.1   Magnitude of motion 158
        8.1.1.1   Patient motion 158
        8.1.1.2   Respiratory motion 158
        8.1.1.3   Cardiac motion 159
8.2   Motion correction on 3D PET data 160
    8.2.1   Overview 161
    8.2.2   Rigid motion correction 162
    8.2.3   Elastic motion correction 163
8.3   Optical flow 164
    8.3.1   Image constraint equation 164
    8.3.2   Optical flow methods 166
    8.3.3   Optical flow in medical imaging 167
8.4   Lucas–Kanade optical flow 168
8.5   Horn–Schunck optical flow 169
8.6   Bruhn optical flow 170
8.7   Preserving discontinuities 172
8.8   Correcting for motion 173
8.9   Mass conservation–based optical flow 174
    8.9.1   Correcting for motion 175
References 177

8.1   Introduction

Motion is a major cause of image distortion and degradation in emission tomography. The foremost effect of motion on the reconstructed images is the induction of blur which is proportional in magnitude to the amount of motion causing it. Small objects or those with low contrast may thus become invisible as the activity is spread over a larger image volume. It has been shown that the respiratory motion may lead to incorrect staging of tumors [25]. Small tumors may even succeed in evading detection [50]. In addition to image blur, another disadvantage of motion has been observed and quantified in recent years on PET images: attenuation artifacts in the case of hybrid scanners where CT- or MRI-based attenuation maps are used for attenuation correction of the PET data.

As the motion correction techniques are similar for SPECT and PET to-mographic images, or else can be easily modified to be applied to the other modality, the correction techniques in this chapter will mainly be described with reference to PET imaging only.

The degrading motion in PET/SPECT studies can be divided into cate-gories in accordance with the source of the motion. Three major sources of motion can be differentiated in emission tomography:

- Body motion, caused by willing or unwilling motion of the patient

- Respiratory motion, caused by the motion of the diaphragm and lungs

- Cardiac motion, caused by the beating motion of the heart

To study the effects and the correction techniques for these types of motion, it would be helpful to first see how large the magnitude of each type of motion is. In most cases, this also gives an idea about the impact of the motion on the images.

8.1.1   Magnitude of motion

8.1.1.1   Patient motion

The motion of the patient may be caused involuntarily by tremor or spasms due to a disease or by motion due to coughing or sneezing by the patient. It may also be caused by the motion of body parts by patients if they become tired or feel pain caused by lying in the same position for long periods of time. The magnitude of this motion is variable. Techniques to minimize the body motion of the patient include usage of a special mechanism such as the BodyFix® system. Other methods include observing the patient during data acquisition and inducing a trigger signal in the data stream whenever large patient movements are observed. The data can then be corrected for motion in a post processing step.

8.1.1.2   Respiratory motion

Respiratory motion of the patients is caused by the movement of the di-aphragm which leads to expansion or contraction of the lungs. Along with the lungs other organs in the thorax, such as the heart, which is located be-tween the lungs and is adjacent to the diaphragm, the liver, the ribs and the

TABLE 8.1: Maximum respiratory motion of different organs.

Organ	Motion [mm]	Reference
Diaphragm	38.0	Vedam et al. [64]
Heart	23.5	McLeish et al. [47]
Liver	25.0	Brandner et al. [6]
Spleen	25.0	Brandner et al.[6]

outer surface of the thorax, the kidneys, etc., are also moved. This respiratory motion is a 3D motion with major components in the cranio-caudal and the anterior-posterior directions and leads to image blur (see Figure 8.1). The respiratory pattern of each patient is unique as can be seen in Figure 8.2.

The maximum motion of different organs due to respiration as measured in clinical studies is given in Table 8.1. It is obvious from the table that the extent of the motion exceeds or is comparable to the dimensions of small objects of interest. This means that in absence of motion correction, the activity from such organs or tumors might be spread over a relatively large volume and distort the quantification and analysis of the data.

8.1.1.3 Cardiac motion

The heart performs a complex motion during its contraction phase: it not only contracts in length but also twists during the contraction. A major change in the size and shape of the heart results. The blood volume inside the heart is thereby reduced and the thickness of the heart muscle, the myocardium,

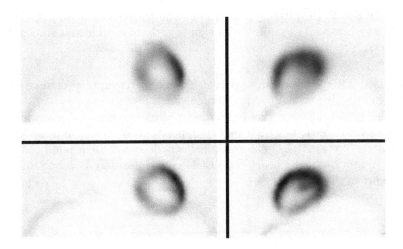

FIGURE 8.1: The respiratory motion causes blurring of PET data. Left: coronal, right: saggital slices of a cardiac study. The effect of image blur is evident from the images. Top row shows all PET data, bottom row shows only one phase from the data.

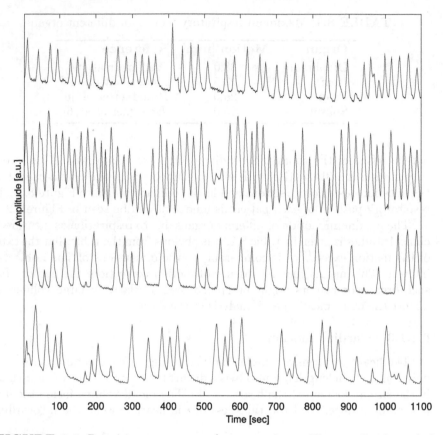

FIGURE 8.2: Respiratory pattern of some patients. The amplitude and the cycle length as well as shape varies significantly with time and among different patients.

increases. This leads to an apparent change in the uptake estimated on the PET images due to the partial volume effect.

The motion of heart due to cardiac motion is estimated at 25 mm [63] in some cases. It must be noted that in opposition to the respiratory signal, the amplitude of the ECG signal from the heart does not correspond to the actual physical motion of the heart. The ECG gives information about the phase of the heart cycle only.

8.2   Motion correction on 3D PET data

The basic idea for motion correction in emission tomography is to correct the data for motion by first estimating the motion and then compensating for it. Several methods have been proposed to this end. These include use of an *a priori* model of motion, discarding the data from some phases of the respiratory or cardiac cycles which show the greatest motion, dividing the emission data into phases and selecting only a particular phase of it, use of registration techniques to deform the data, etc.

As the respiratory motion is continuous, it is possible to divide the emission data into several parts, each with minimal motion[17]. This is called gating. Gating would lead to reducing the amount of motion in each phase; however, it would also lead to an increase in noise on the reconstructed images due to less information present in each phase. Therefore, selecting only the best phase is not the complete solution to the problem at hand. A deformation of the complete gated data to the target phase is required. This will allow the full information along with less motion.

A good method for motion correction would be applicable to the PET data without any prior segmentation or assumptions on the properties of the different types of tissue present in the body. Before describing the motion correction techniques, a very short overview of the past attempts at motion correction in emission tomographic studies is given.

8.2.1 Overview

In those parts of the body where only rigid motion is present, i.e., translation and rotation, simpler image registration algorithms can be used. One such set of clinical studies are the brain studies. Picard et al. [52] and Fulton et al. [28] used video cameras to monitor the skull position externally and correct for the motion in a post processing step by registration. More recent attempts also use external monitoring [10] to correct the lines of responses (LORs) for the motion before reconstruction.

In some studies external motion estimation is used for sorting the events in accordance with an external motion signal; e.g., Nehmeh et al. [48] used an external block fastened to the patient's abdomen. The movement of this block is tracked with video cameras and is used for sorting the data into different phases. Only the first phase or gate is used for image reconstruction. This type of method reduces the motion on the images; however, most of the information is lost in the unused gates. To get the same statistics as in the ungated images, a proportionally larger amount of the radiotracer must be used or the acquisition time has to be prolonged accordingly. Nehmeh et al. also proposed another method [49] based on the use of an external radioactive point source which is placed inside the field of view. The emission data is then acquired in multiple short duration frames. All frames in which the point source falls within a region of interest (ROI) around the point source are summed up and used for the reconstruction. This method utilizes more information than the previous one but still a large part of the events is lost. These methods cannot be strictly called motion correction; it is rather a selection of frames which are "good."

Klein et al. [35] used deformable elastic membranes to model the cardiac motion in heart studies successfully. However, this method is applicable to the heart only as it models the elasticity of the myocardium and requires prior segmentation of the heart. As the underlying properties of the elastic membranes change for each type of tissue and fluid, different models are required for a comprehensive motion correction of the thorax. A similar method was proposed by Zhang et al. [67]. The performance of these methods depends upon the accuracy of the parameters derived from tests with the tissue. The methods are applicable in 2D and prior global registration is also required.

Some studies attempted motion correction in the pre-image reconstruction stage, e.g., in sinograms [39] or list mode [10]. High intensity nodes (if they are present in the field of view) can be used to correct the inplane motion on sinograms, whereas external tracking with markers can be used to re-sort the events in accordance with the respiratory phase.

Studies that estimate the motion on gated CT images, e.g., by Manjeshwar[45], Qiao[55], Mair [44], are not of further interest for this brief survey because they lead to an increase in radiation which is not justifiable for most routine patients.

8.2.2   Rigid motion correction

It has been already mentioned that some of the motion on emission tomographic images is caused by rigid motion by the patient, such as the movement of the head in brain studies. Rigid registration algorithms, which correct for the translational and rotational motion, are useful in this scenario. The task is to find parameters using a given (or floating) image that can be brought into spatial correspondence or alignment with a target image.

If both the floating and the target images belong to the same modality, e.g., PET, the registration task can be expressed mathematically as

$$I_T(x,y,z) = I_F(f(x,y,z)) \qquad (8.1)$$

where x, y and z are the coordinates of the 3D image volume along the corresponding axis, I_T is the target image, I_F is the floating image and f is a coordinate transformation function.

In rigid registration usually 3 parameters for rotation and 3 parameters for translation are used. Calling α, β, γ as rotation and t_x, t_y, t_z as translation parameters, the rigid transformation can then be given as

$$f_r = A\vec{v}$$

where $A =$

$$\begin{pmatrix} \cos\beta\cos\gamma & \cos\alpha\sin\gamma+\sin\alpha\sin\beta\cos\gamma & \sin\alpha\sin\gamma-\cos\alpha\sin\beta\cos\gamma & t_x \\ -\cos\beta\sin\gamma & \cos\alpha\cos\gamma-\sin\alpha\sin\beta\sin\gamma & \sin\alpha\cos\gamma+\cos\alpha\sin\beta\sin\gamma & t_y \\ \sin\beta & -\sin\alpha\cos\beta & \cos\alpha\cos\beta & t_z \\ 0 & 0 & 0 & 1 \end{pmatrix}$$

and

$$\vec{v} = \begin{pmatrix} x \\ y \\ z \\ 1 \end{pmatrix}.$$

Finding the unknown parameters is now the task, which can be solved by iteratively applying a set of parameter values and calculating the distance between the newly transformed floating image and the target image. The parameter values are then adjusted after each iteration as long as some stopping criterion is not fulfilled. Many distance measures and stopping criteria are possible, some of which have been described in [7],[42],[43],[53],[69]. Using special distance measures that are applicable to more than one modality simultaneously, registration between multi-modal images can also be achieved. An example of rigid registration is shown in Figure 8.3.

In rigid registration the angles between the lines are preserved; i.e., a square will remain a square after rigid registration. The method fails if the deformed (floating) image has not preserved the angles between the lines. To deal with such motion affine registration methods are required. These methods use additionally 3 parameters for scaling and 3 for the skew. With this, a square can be registered to a parallelogram.

8.2.3 Elastic motion correction

The rigid or affine registration methods provide a global transformation function for the whole image. However, in medical imaging the motion of the organs, especially in the thorax, does not follow a global pattern. The organs describe different forms of motion and undergo deformations. For example, the lungs expand in all directions during the inspiration, whereas the heart is pressed downwards during the inhalation. The motion is not only in different directions but also with varying magnitudes. Rigid or affine registration methods are thus not sufficient and are inferior to non-rigid registration methods in terms of accuracy [19]. The elastic registration methods describe the deformation between the target and the floating image with the help of a vector

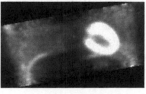

(a) Target image   (b) Floating image

FIGURE 8.3: Rigid motion. If the motion consists of only rotations and translations, it is called rigid motion and is easy to correct.

field over the whole image volume. Thus a deformation function for each voxel is estimated. This means that a line in one image volume can be registered to a curve in the other.

The elastic registration algorithms can be classified according to the underlying model that they use to estimate the deformations. Following this scheme of classification, there are three main categories of such algorithms. These are based on:

- physical models

- interpolation methods

- transformation constraints

The algorithms of the first category use such physical models as stress-strain deformation or fluid flow equations. Optical flow algorithms, which will be discussed in detail here, also belong to this category. Algorithms of the second category use basis functions such as splines, polynomials or wavelets whose coefficients are adjusted to give an approximation of the displacement field. The third category includes methods that use transformation constraints. These include the transformation consistency; i.e., the registration of A with B should correspond to registration of B with A or the transformation should be bijective, etc. These methods may actually use one of the algorithms of the first two categories and add a constraint to it.

Simple elastic registration methods show disadvantages in the case of localized deformations [69]. Interpolation- and approximation-based algorithms such as the splines based methods encounter problems at finer resolution scales [13] in addition to the fact that they do not model any physical or biological process. The optical flow methods allow a higher degree of freedom in estimating the motion of different organs simultaneously and model the motion of a fluid.

8.3   Optical flow

Optical flow is the change in the intensity pattern of the voxels between two image volumes. A vector field, called "flow field," can be estimated by modeling the voxel intensity differences between the two image volumes on the pattern of the flow of a fluid (see Figure 8.4).

8.3.1   Image constraint equation

If two image volumes are given at times t and $t + \delta t$. A voxel at location (x, y, z, t) with intensity $I(x, y, z, t)$ will move by $\delta x, \delta y, \delta z$ and δt between the

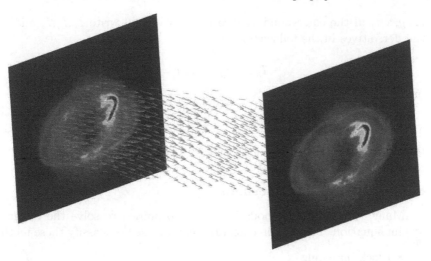

FIGURE 8.4: The optical flow is modeled as smooth flow of a fluid.

two volumes. Therefore the following image constraint equation (also called brightness consistency constraint) holds true:

$$I(x, y, z, t) = I(x + \delta x, y + \delta y, z + \delta z, t + \delta t). \qquad (8.2)$$

Assuming the movement to be small enough, an image constraint can now be developed from Equation (8.2) at $I(x, y, z, t)$ with Taylor series:

$$
\begin{aligned}
I(x + \delta x, y + \delta y, z + \delta z, t + \delta t) \quad = \quad & I(x, y, z, t) + \frac{\partial I}{\partial x}\delta x + \frac{\partial I}{\partial y}\delta y + \frac{\partial I}{\partial z}\delta z \\
& + \frac{\partial I}{\partial t}\delta t + H
\end{aligned}
$$

where H means higher order terms, which should be small enough to be ignored. From these equations we achieve:

$$\frac{\partial I}{\partial x}\delta x + \frac{\partial I}{\partial y}\delta y + \frac{\partial I}{\partial z}\delta z + \frac{\partial I}{\partial t}\delta t = 0$$

or

$$\frac{\partial I}{\partial x}\frac{\delta x}{\delta t} + \frac{\partial I}{\partial y}\frac{\delta y}{\delta t} + \frac{\partial I}{\partial z}\frac{\delta z}{\delta t} + \frac{\partial I}{\partial t}\frac{\delta t}{\delta t} = 0$$

which results in

$$\frac{\partial I}{\partial x}u + \frac{\partial I}{\partial y}v + \frac{\partial I}{\partial z}w + \frac{\partial I}{\partial t} = 0$$

where u, v, w are the x, y and z components of the velocity $\vec{V}$ or optical flow of $I(x, y, z, t)$, and $\frac{\partial I}{\partial x}, \frac{\partial I}{\partial y}, \frac{\partial I}{\partial z}$ and $\frac{\partial I}{\partial t}$ are the derivatives of the image at

(x, y, z, t) in the corresponding directions. We shall write I_x, I_y, I_z and I_t for the derivatives in the following.

Thus

$$I_x u + I_y v + I_z w = -I_t$$

or

$$\nabla I \cdot \vec{V} = -I_t. \tag{8.3}$$

This is a single equation with three unknowns and therefore the system is under-determined. To find the optical flow additional equations are required.

8.3.2   Optical flow methods

Many optical flow methods have been proposed to solve the image constraint equation. Three main categories can be used to classify these methods:

- Block matching

- Frequency domain correlation

- Gradient based

In the block matching–based optical flow algorithms, small blocks of one image are moved so that they match with a block on the other image. The motion vector gives the transformation for this block only. All blocks in an image are moved until vectors are found that minimize some error function. It was shown by Davis/Freeman that block matching optical flow algorithms are equivalent to gradient-based algorithms if the displacements are sub-voxel and the deformations are rigid body deformations [14]. Zhang/Lu have even combined both approaches to a hybrid block matching algorithm which utilizes gradient information [66]. Behar et al. combined block matching with gradient-based optical flow in a small window [5]. Correlation-based optical flow as presented by Duan et al. [23] is also a form of block matching.

One way of detecting motion on the images is to analyze their frequency transforms such as the Fourier transform. The difficulty with such an approach is that the Fourier transform is global and thus scenes including many objects undergoing motion render it not suitable for this task. Reed [57] uses a Gabor filter–based local frequency approach to overcome this problem and detect motion in the frequency domain of the images. A similar approach was already used by Fleet/Japson [26]. They use a decomposition of the images into band-pass channels and use a phase constraint equation in each channel to detect motion. Prince et al. use a combination of band-pass and gradient based optical flow [54]. Phase correlation-based optical flow [61] also falls under this category.

There have been many proposals for image gradient–based methods of estimating optical flow. Lucas–Kanade proposed a solution that assumed the flow to be constant in a local neighborhood around the central voxel and used the least means of square approach [40]. Bab-Hadiashar improved this method by taking a least median of squares approach at solving the over-determined linear equation system [2]. Horn–Schunck assumed a global smoothing function to solve the image constraint equation[34]. This method was improved by El-Feghali/Mitchie to preserve boundaries by weighting the smoothing term [24]. Deriche et al improved the Horn–Schunck algorithm by using a non-quadratic function for smoothing [20]. A different method was proposed by Alvarez et al. [1] who improved the algorithm by Nagel/Enkelmann by using a brightness invariant term into consideration. Thus if two images differ in contrast by a constant factor, their method can still estimate the flow correctly. The local and global methods were combined by Vemuri et al. to register medical images [65]. Lastly, Bruhn et al. [8] have given a framework for combined local and global optical flow algorithms.

Other approaches to optical flow estimation include the mass conservation method [56],[59], fluid mechanics–based method [12], or approaches using higher order constraints on flow [27].

The gradient-based algorithms for optical flow estimation achieve high accuracy and flexibility. It is thus no wonder that most of the research has been done in the area of gradient-based methods. The image volumes reconstructed from the PET data after appropriate gating contain equal amounts of radioactivity. As all gates contain data from almost all breathing cycles, the basic image constraint equation is fulfilled in the PET data. Thus the gradient-based methods can be applied to our problem readily.

8.3.3   Optical flow in medical imaging

Medical imaging has also been a fertile field for the use of optical flow methods. Most of the applications in this respect are concerned with motion detection, registration and image segmentation on medical images. Modalities that have benefited from optical flow applications include, but are not confined to, magnetic resonance imaging (MRI), computed tomography (CT), positron emission tomography (PET) and ultrasound (US) imaging.

Hata et al. used optical flow to measure the deformations between MRI images of the brain taken before, during and after brain surgery [33]. Zientara et al. used optical flow to measure dilation and contraction of liver tissue during laser ablation on MRI images [68]. Another application was to estimating cardiac displacements in tagged MRI images [21],[54]. Optical flow has been used to segment heart on gated MRI images by Galic et al. [29]. Recently successful tracking of endocardium on MRI images was shown by Duan et al. [22].

On the CT side, Song et al. [59] and Gorce et al. [32] used optical flow to estimate 3D velocity fields on CT cardiac images. A study by Torigian et al. suggests use of optical flow for assessing regional air trapping in lung CT images acquired in inspiration and expiration phases [62]. A method for prospective motion correction of X-ray imaging of the heart was presented by Shechter et al. [58].

Klein [35] was probably one of the first to apply optical flow to cardiac PET studies for motion correction. Vemuri et al. used optical flow to register medical images [65]. Dawood et al. have used optical flow to correct PET images for the effects of motion [17],[16],[18].

Among other medical imaging modalities, Behar et al. used optical flow in echocardiography [5].

We now turn to three optical flow algorithms based on the image gradients.

8.4   Lucas–Kanade optical flow

Lucas and Kanade use a non-iterative method which that a locally constant flow [40]. Evaluations and comparisons by Barron [3], Bruhn [9] and Galvin [30] have shown that the Lucas–Kanade algorithm is one of the best methods for calculating optical flow fields especially in presence of noise.

Assuming that the flow (u, v, w) is constant in a small window of size $m \times m \times m$ with $m > 1$, which is centered at voxel x, y, z and numbering the pixels as $1...n$ we get a set of equations:

$$I_{x_1} u + I_{y_1} v + I_{z_1} w = -I_{t_1}$$
$$I_{x_2} u + I_{y_2} v + I_{z_2} w = -I_{t_2}$$
$$\vdots \qquad \vdots \qquad \vdots$$
$$I_{x_n} u + I_{y_n} v + I_{z_n} w = -I_{t_n}.$$

With this we get an over-determined system:

$$
\begin{pmatrix}
I_{x_1} & I_{y_1} & I_{z_1} \\
I_{x_2} & I_{y_2} & I_{z_2} \\
\vdots & \vdots & \vdots \\
I_{x_n} & I_{y_n} & I_{z_n}
\end{pmatrix}
\begin{pmatrix}
u \\
v \\
w
\end{pmatrix}
=
\begin{pmatrix}
-I_{t_1} \\
-I_{t_2} \\
\vdots \\
-I_{t_n}
\end{pmatrix}
$$

or

$$A\vec{V} = -b.$$

To solve the this system of equations the least squares method can be used:

$$A^T A \vec{V} = A^T(-b)$$

or

$$\vec{V} = (A^T A)^{-1} A^T(-b) \tag{8.4}$$

or

$$
\begin{pmatrix}
u \\
v \\
w
\end{pmatrix}
=
\begin{pmatrix}
I_x^2 & I_x I_y & I_x I_z \\
I_x I_y & I_y^2 & I_y I_z \\
I_x I_z & I_y I_z & I_z^2
\end{pmatrix}^{-1}
(-A^T I_t).
$$

A weighting function $W(i,j,k)$, with $i,j,k \in [1,..,m]$ can be used to give more prominence to the center voxel of the window. One of the characteristics of the Lucas–Kanade algorithm is that it does not yield a high density of flow vectors; i.e. the flow information fades out quickly across motion boundaries. The main advantage of this method is its comparative robustness in presence of noise [3],[8].

8.5 Horn–Schunck optical flow

The Horn–Schunck optical flow method introduces the additionally required constraint by postulating a global condition. It assumes a global smoothness in the flow field. This method was applied to medical imaging by Song et al. [59] to calculate the flow field in CT images of the heart. The energy function to be minimized is given by combining the image constraint (see Equation 8.2) with a smoothing functional.

$$f = \int ((\nabla I \cdot \vec{V} + I_t)^2 + \alpha^2(|\nabla u|^2 + |\nabla v|^2 + |\nabla w|^2))\mathrm{d}x\mathrm{d}y\mathrm{d}z \qquad (8.5)$$

The parameter α is a regularization constant. Larger values of α lead to a smoother flow. This equation can be solved by calculating the corresponding Euler–Lagrange equations which are given as

$$\Delta u - \frac{1}{\alpha^2}I_x(I_x u + I_y v + I_z w + I_t) = 0$$

$$\Delta v - \frac{1}{\alpha^2}I_y(I_x u + I_y v + I_z w + I_t) = 0 \qquad (8.6)$$

$$\Delta w - \frac{1}{\alpha^2}I_z(I_x u + I_y v + I_z w + I_t) = 0$$

where Δ denotes the Laplace operator so that

$$\Delta = \nabla^2 = \frac{\partial^2}{\partial x^2} + \frac{\partial^2}{\partial y^2} + \frac{\partial^2}{\partial z^2}.$$

Using the definition of the Laplacian operator in image processing: $\Delta u = \bar{u} - u$, where $\bar{u}$ is the average of u in the neighborhood of the current voxel position we arrive at the equation system:

$$\begin{pmatrix} \alpha^2 + I_x^2 & I_x I_y & I_x I_z \\ I_x I_y & \alpha^2 + I_y^2 & I_y I_z \\ I_x I_z & I_y I_z & \alpha^2 + I_z^2 \end{pmatrix} \begin{pmatrix} u \\ v \\ w \end{pmatrix} = \begin{pmatrix} \alpha^2 \bar{u} - I_x I_t \\ \alpha^2 \bar{v} - I_y I_t \\ \alpha^2 \bar{w} - I_z I_t \end{pmatrix}.$$

Solving this system for the variables

$$u^{k+1} = \overline{u^k} - \frac{I_x(I_x\overline{u^k} + I_y\overline{v^k} + I_z\overline{w^k} + I_t)}{\alpha^2 + I_x^2 + I_y^2 + I_z^2}$$

$$v^{k+1} = \overline{v^k} - \frac{I_y(I_x\overline{u^k} + I_y\overline{v^k} + I_z\overline{w^k} + I_t)}{\alpha^2 + I_x^2 + I_y^2 + I_z^2} \qquad (8.7)$$

$$w^{k+1} = \overline{w^k} - \frac{I_z(I_x\overline{u^k} + I_y\overline{v^k} + I_z\overline{w^k} + I_t)}{\alpha^2 + I_x^2 + I_y^2 + I_z^2}$$

where the superscript k denotes the iteration number.

Advantages of the Horn–Schunck algorithm include that it yields a high density of flow vectors; i.e., the flow information missing in inner parts of homogeneous objects is *filled in* from the motion boundaries. However, it is more sensitive to noise than local methods [8],[46].

8.6   Bruhn optical flow

It is obviously advantageous to combine both types of algorithms to get an algorithm which is robust in presence of noise and yields dense vector fields. Bruhn [8] has presented a mathematical framework which allows combinations of both Lucas–Kanade as well as Horn–Schunck algorithms. Defining:

$$V = \begin{pmatrix} u \\ v \\ w \\ 1 \end{pmatrix} \quad , \quad \nabla I = \begin{pmatrix} I_x \\ I_y \\ I_z \\ I_t \end{pmatrix} \text{ and}$$

$$|\nabla V|^2 = |\nabla u|^2 + |\nabla v|^2 + |\nabla w|^2$$

we can rewrite the Lucas–Kanade algorithm as minimization of a function f_{LK}:

$$\begin{aligned} f_{LK} &= V^T(k * (\nabla I \nabla I^T))V \\ &= k * (I_x u + I_y v + I_z w + I_t)^2 \end{aligned} \qquad (8.8)$$

where k is a weighting function which is convolved with the image data. Minimization of Equation (8.8) means $\partial_u f_{LK} = 0$, $\partial_v f_{LK} = 0$, $\partial_w f_{LK} = 0$.

Similarly the Horn–Schunck can be rewritten as minimization of a function f_{HS}:

$$f_{HS} = \int (V^T \nabla I \nabla I^T V + \alpha |\nabla V|^2) dx dy dz. \qquad (8.9)$$

Combining both functions gives us a hybrid local-global function:

$$f_{LG} = \int (k * V^T(\nabla I \nabla I^T)V + \alpha|\nabla V|^2)\mathrm{d}x\mathrm{d}y\mathrm{d}z. \tag{8.10}$$

To avoid heavy penalization of outliers a square root function can be used

$$\psi_i(s^2) = 2\beta_i^2\sqrt{1 + \frac{s^2}{\beta_i^2}}, i \in 1,2 \tag{8.11}$$

where β_i is a scaling factor [8]. With this improvement the function f_{LG} becomes convex in s and thus has a unique solution:

$$\mathrm{div}(\psi_2'(S)\nabla u) - \frac{1}{\alpha}\psi_1'(D)I_x(I_xu + I_yv + I_zw + I_t) = 0$$

$$\mathrm{div}(\psi_2'(S)\nabla v) - \frac{1}{\alpha}\psi_1'(D)I_y(I_xu + I_yv + I_zw + I_t) = 0$$

$$\mathrm{div}(\psi_2'(S)\nabla w) - \frac{1}{\alpha}\psi_1'(D)I_z(I_xu + I_yv + I_zw + I_t) = 0$$

with $D = k * V^T(\nabla I \nabla I^T)V$ and $S = |\nabla u|^2 + |\nabla v|^2 + |\nabla w|^2$.
The algorithmic solution is then given by:

$$u_i^{k+1} = \frac{\sum\limits_{j\in N} \frac{\psi_{2i}'+\psi_{2j}'}{2}u_{x_j}^k - \psi_{1i}'\frac{h^2}{\alpha}I_{x_i}\left(I_{y_i}v_i^k+I_{z_i}w_i^k+I_{t_i}\right)}{\sum\limits_{j\in N} \frac{\psi_{2i}'+\psi_{2j}'}{2}+\psi_{1i}'\frac{h^2}{\alpha}I_{x_i}^2}$$

$$v_i^{k+1} = \frac{\sum\limits_{j\in N} \frac{\psi_{2i}'+\psi_{2j}'}{2}v_{y_j}^k - \psi_{1i}'\frac{h^2}{\alpha}I_{y_i}\left(I_{x_i}u_i^k+I_{z_i}w_i^k+I_{t_i}\right)}{\sum\limits_{j\in N} \frac{\psi_{2i}'+\psi_{2j}'}{2}+\psi_{1i}'\frac{h^2}{\alpha}I_{y_i}^2}$$

$$w_i^{k+1} = \frac{\sum\limits_{j\in N} \frac{\psi_{2i}'+\psi_{2j}'}{2}w_{z_j}^k - \psi_{1i}'\frac{h^2}{\alpha}I_{z_i}\left(I_{x_i}u_i^k+I_{y_i}v_i^k+I_{t_i}\right)}{\sum\limits_{j\in N} \frac{\psi_{2i}'+\psi_{2j}'}{2}+\psi_{1i}'\frac{h^2}{\alpha}I_{z_i}^2}$$

with

$$\psi_{1i}'(D) = \frac{1}{\sqrt{1 + \frac{(I_{x_i}u_i+I_{y_i}v_i+I_{z_i}w_i+I_{t_i})^2}{\beta_i^2}}} \tag{8.12}$$

and

$$\psi_{2i}'(S) = \frac{1}{\sqrt{1 + \frac{|\nabla u_i|^2+|\nabla v_i|^2+|\nabla w_i|^2}{\beta_i^2}}} \tag{8.13}$$

where i is the current pixel position.

The updated elements of the flow should be immediately used for the next step, as required by the Gauss–Seidel method. Thus only one variable array is needed for the calculation which is constantly updated.

8.7   Preserving discontinuities

Smoothing ensures high density in the optical flow but it has the negative impact of fading out the flow across the boundaries of moving objects. Physiologically the organs in the human body perform different types of motion but the tissue of one organ never flows inside another. Thus it is necessary to accentuate the organ boundaries, which amounts to preserving discontinuities in the flow.

A whole range of methods has been proposed for preserving these discontinuities. The techniques include using variable values for α or selecting only a part of equations in the case of local methods [4] or to use a function that tries to preserve discontinuities by weighting the smoothing term [8], [20], [24].

An appropriate weighting function should have the property of reducing the effect of the smoothing term in areas of discontinuity, i.e., at object boundaries, but to retain the effect in areas of homogeneous flow such as inside the objects. We have adopted a function by weighting the smoothing term in accordance with the image gradient. Thus places where edges are present will be smoothed less than areas inside organs which usually have fewer edges and a lower gradient. This in turn leads to the reformulation of equation (8.13) as

$$\psi'_{2i}(S) = \frac{1}{\sqrt{1 + \frac{|I_x|*|\nabla u_i|^2 + |I_y|*|\nabla v_i|^2 + |I_z|*|\nabla w_i|^2)}{\beta_i^2}}}. \tag{8.14}$$

The discontinuity-preserving algorithm uses coupled pairs of equations. The optical flow is calculated iteratively in a double loop. The outer loop updates the ψ variables and the inner loop updates the u, v, w optical flow components. The algorithm in pseudo code can be described as follows:

```
initialize u, v, w, psi_1, psi_2
for outer loop
  for x,y,z
    update psi_1'    as in equation 1.12
    update psi_2'    as in equation 1.15
  end
  for inner loop
    for x,y,z
      update u
```

```
      update v
      update w
    end
  end
end
```

8.8   Correcting for motion

Once the motion is estimated, it can be corrected by inverting the motion vectors at every pixel position and deforming the 3D image according to the motion vector field. As the motion vectors obtained from the optical flow algorithms are not always integers, interpolation has to be employed [41],[37],[60],[38],[51]. A tri-linear interpolation method can be used. All motion-corrected gates can then be averaged to get the PET data with minimal motion (Figure 8.5).

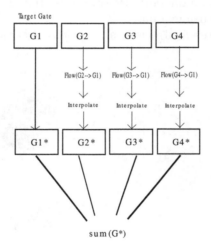

FIGURE 8.5: The process flow (eight gates were actually used; only four are shown here for simplicity).

8.9   Mass conservation–based optical flow

The above mentioned methods for motion correction are applicable to the PET data only when correspondences in the gray values of the target and floating images are given. In cardiac motion, this is not the case. The so-called partial volume effect (PVE) becomes prominent in cardiac studies. It is a result of the limited resolution of the scanners. All objects smaller than the scanner resolution limit cannot be accurately delimited and therefore appear blurred. As the heart muscle contracts and expands during the cardiac cycle its thickness varies. In phases with thicker heart wall, the activity is better resolved and has a higher amplitude as compared to other phases where the myocardium is thin (see Figure 8.6). However, the total amount of the activity remains the same.

Two important studies related to the correction for cardiac motion are [36] and [31]. In the first study optical flow is used for estimating the deformations in the images by modeling the myocardium as an elastic membrane. The second study combines the motion estimation of the first study with reconstruction in a single framework. However this study is confined to 2D images and deals with cardiac motion.

The method presented below is essentially different from the brightness

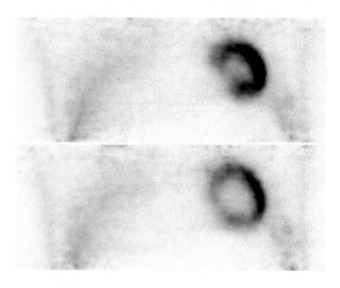

FIGURE 8.6: Two phases from the cardiac cycle of the heart. Above: end-systole, Below: end-diastole. A coronal slice from the 3D PET image volume is shown. Images from an FDG study are shown here without attenuation correction.

consistency-based optical flow methods [16],[8] described before as it is based on the continuity equation. This change in the basic model is necessary as brightness consistency is not given in cardiac-gated PET data due to the PVE.

The law of mass conservation says that the total mass in a closed system is conserved. If we substitute activity by mass, the law must still hold in our case, as the total activity in the system remains the same from systolic to diastolic phases of the heart. It is blurred only in the diastolic phase. It should be noted that our data is pre-corrected for the time-dependent radioactive decay during the reconstruction process so that the decay itself plays no role in our considerations.

The continuity equation for mass conservation is given as [11]

$$\frac{\partial I}{\partial t} + div(I\mathbf{u}) = 0 \tag{8.15}$$

where I is the intensity value, $\mathbf{u} = (u, v, w)^T$ is the velocity vector, i.e., the optical flow. Deviations from this equation can be penalized by the following functional:

$$\int (\nabla I \cdot \mathbf{u} + I_t + I \cdot div(\mathbf{u}))^2 \mathrm{d}x\mathrm{d}y\mathrm{d}z. \tag{8.16}$$

The derivative in time I_t can be calculated on discrete image volumes by using the difference: $I_2 - I_1$, where I_2 is the floating and I_1 is the target image volume. As with the intensity-based optical flow, this is again an under-determined system of equations and therefore a smoothing term can be added to solve it. The same smoothing term as given in the Horn–Schunck algorithm above can be used here as well. The resulting optical flow functional is thus:

$$f \quad = \quad \mathrm{argmin}\left[\int_V D^2 dV + \alpha \int_V SdV\right] \tag{8.17}$$

with

$$D \quad = \quad div(I\mathbf{u}) + I_t, \quad S = |\nabla u|^2 + |\nabla v|^2 + |\nabla w|^2.$$

The minimization of the equation (8.17) can be achieved by using the corresponding Euler–Lagrange equations. These are given by

$$\begin{aligned} 0 &= D_x I + \alpha \Delta u \\ 0 &= D_y I + \alpha \Delta v \\ 0 &= D_z I + \alpha \Delta w \end{aligned} \tag{8.18}$$

where D_x, D_y, D_z are the first derivatives of D in the corresponding directions. Here, α is a weighting parameter.

8.9.1 Correcting for motion

Once the optical flow is found (see Figure 8.7) the images have to be transformed to get the motion-corrected data. Equation (8.15) can be used

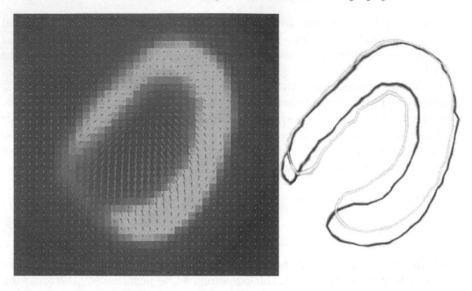

FIGURE 8.7: The optical flow calculated with the proposed method. A coronal slice is shown with superimposed vectors. Only two components of the flow are shown.

for this purpose. As the time derivative I_t was calculated as $I_2 - I_1$ and the flow $\mathbf{u}$ is now assumed to be known, the transformed image can be calculated as

$$I_{2mc} = I_2 + div(I_2\mathbf{u}). \tag{8.19}$$

Applying this equation, the deformed (motion corrected) images can be readily calculated from the floating images.

The motion correction methods described above have been applied to real patient data and their performance measured with various methods. Different aspects such as noise levels, extent of motion, pattern of breathing, etc. were taken into account, too. The results showed that the discontinuity preserving method performed well on real patient data. The respiratory motion was detected and correspondingly corrected to well below voxel size levels [15][16].

An example of the application of the brightness consistency based optical flow applied to PET data is shown in Figure 8.8. The left column shows, from top to bottom, the target phase, the floating phase and the floating phase after motion correction, respectively. The white line on the images helps assess the position of the heart and diaphragm. It shows that the heart and the diaphragm are displaced downwards and a little to the left (on images) as the patient inhales from target to floating phases. The motion-corrected version has all image parts in the same position as in the target phase. In the right column of the figure half of the estimated flow vectors are shown.

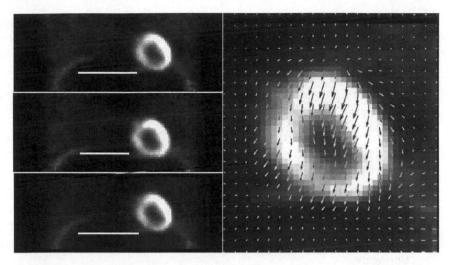

FIGURE 8.8: Optical flow applied to real patient data. The left column shows from top to bottom: Target phase, floating phase and the floating phase after motion correction. The right column shows the estimated flow vectors. Only coronal slices are shown. Similarly, only half of the vectors are shown for better visibility.

References

[1] L. Alvarez, J. Weickert, and J. Sanchez. Reliable estimation of dense optical flow fields with large displacements. *International Journal of Computer Vision*, 39(1):41–56, 2000.

[2] A. Bab-Hadiashar and D. Suter. Robust total least squares based optic flow computation. In *Asian Conference on Computer Vision*, volume 1, 1998.

[3] J.L. Barron, D.J. Fleet, and S.S. Beauchemin. Performance of optical flow techniques. *International Journal of Computer Vision*, 12:43–77, 1994.

[4] F. Bartolini, V. Capellini, C. Colombo, and A. Mecocci. A multiwindow least squares approach to the estimation of optical flow with discontinuities. *Optical Engineering, Special Section, From Numerical to Symbolic Image Processing: Systems and Applications*, 32(6):1250–1256, 1993.

[5] V. Behar, D. Adam, P. Lysyansky, and Z. Friedman. Improving motion estimation by accounting for local image distortion. *Ultrasonics*, 43(1):57–65, Oct 2004.

[6] E.D. Brandner, A. Wu, H. Chen, D. Heron, S. Kalnicki, K. Komanduri, K. Gerszten, S. Burton, I. Ahmed, and Z. Shou. Abdominal organ motion measured using 4D CT. *International Journal of Radiation Oncology Biology Physics*, 65(2):554–560, Jun 2006.

[7] L.G. Brown. A survey of image registration techniques. *ACM Computing Surveys*, 24:325–376, 1992.

[8] A. Bruhn, J. Weickert, C. Schnörr. Lucas–Kanade meets Horn–Schunck: combining local and global optic flow methods. *International Journal of Computer Vision*, 61(3):211–231, 2005.

[9] A. Bruhn, J. Weickert, and C. Schnörr. Combining the advantages of local and global optic flow methods. In Luc van Gool, editor, *Proceedings of the 24th DAGM Symposium on Pattern Recognition*, volume 2449 of *Lecture Notes on Computer Science*, pages 454–462, Heidelberg, 2002. Springer.

[10] P. Bühler, U. Just, E. Will, J. Kotzerke, and J. van den Hoff. An accurate method for correction of head movement in PET. *IEEE Transactions on Medical Imaging*, 23(9):1176–1185, Sep 2004.

[11] T. Corpetti, D. Heitz, G. Arroyo, E. Mémin, and A. Santa-Cruz. Fluid experimental flow estimation based on an optical-flow scheme. *Experiments in Fluids*, 40:80–97, 2006.

[12] T. Corpetti, E. Memin, and P. Perez. Estimating fluid optical flow. In *Proceedings of the 15th International Conference on Pattern Recognition*, volume 3, page 7045ff, 2000.

[13] W.R. Crum, D. Phil, T. Hartkens, and D.L.G. Hill. Non-rigid image registration: Theory and practice. *British Journal of Radiology*, 77, S140–S153, 2004.

[14] C.Q. Davis and D.M. Freeman. Equivalence of subpixel motion estimators based on optical flow and block matching. In *ISCV '95: Proceedings of the International Symposium on Computer Vision*, page 7ff, 1995.

[15] M. Dawood, C. Brune, F. Büther, X. Jiang, M. Burger, O. Schober, M. Schäfers, and K.P. Schäfers. A Continuity Equation Based Optical Flow Method for Cardiac Motion Correction in 3D PET Data. *Proceedings of the MICCAI 2010 LNCS*, 6326: 88–97, 2010.

[16] M. Dawood, F. Büther, X. Jiang, and K.P. Schäfers. Respiratory motion correction in 3-D PET data with advanced optical flow algorithms. *IEEE Transactions on Medical Imaging*, 27(8):1164–1175, August 2008.

[17] M. Dawood, F. Büther, N. Lang, O. Schober, and K.P. Schäfers. Respiratory gating in positron emission tomography: A quantitative comparision of different gating schemes. *Medical Physics*, 34(7):3067–3076, July 2007.

[18] M. Dawood, N. Lang, X. Jiang, and K.P. Schäfers. Lung motion correction on respiratory gated 3-D PET/CT images. *IEEE Transactions on Medical Imaging*, 25(4):476–485, Apr 2006.

[19] E.R. Denton, L.I. Sonoda, D. Rueckert, S.C. Rankin, C. Hayes, M.O. Leach, D.L. Hill, and D.J. Hawkes. Comparison and evaluation of rigid, affine, and nonrigid registration of breast MR images. *Journal of Computer Assisted Tomography*, 23(5):800–805, 1999.

[20] R. Deriche, P. Kornprobst, and G. Aubert. Optical-flow estimation while preserving its discontinuities: A variational approach. In *Proceedings of the 2nd Asian Conference on Computer Vision*, volume 2, pages 71–80, Singapore, 1995.

[21] L. Dougherty, J.C. Asmuth, A.S. Blom, L. Axel, and R. Kumar. Validation of an optical flow method for tag displacement estimation. *IEEE Transactions on Medical Imaging*, 18(4):359–363, Apr 1999.

[22] Q. Duan, E. Angelini, S. Homma, and A. Laine. Tracking endocardium using optical flow along iso-value curve. *Conf Proc IEEE Engineering in Medical Biology Society*, 1:707–710, 2006.

[23] Q. Duan, E.D. Angelini, S.L. Herz, O. Gerard, P. Allain, C.M. Ingrassia, K.D. Costa, J.W. Holmes, S. Homma, and A.F. Laine. Tracking of LV endocardial surface on real-time three-dimensional ultrasound with optical flow. In AF Farangi et al, editor, *Functional Imaging and Modeling of the Heart 2005*, volume 3504 of *Lecture Notes in Computer Science*, pages 434–445, Berlin, 2005. Springer.

[24] R. El-Feghali and A. Mitiche. Fast computation of a boundary preserving estimate of optical flow. In *Proceedings of the British Machine Vision Conference*, Bristol, UK, 2000.

[25] Y.E. Erdi, S.A. Nehmeh, T. Pan, A. Pevsner, K.E. Rosenzweig, G. Mageras, E.D. Yorke, H. Schoder, W. Hsiao, O.D. Squire, P. Vernon, J.B. Ashman, H. Mostafavi, S.M. Larson, and J.L. Humm. The CT motion quantitation of lung lesions and its impact on PET-measured SUVs. *Journal of Nuclear Medicine*, 45(8):1287–1292, Aug 2004.

[26] D.J. Fleet and A.D. Jepson. Computation of component image velocity from local phase information. *International Journal of Computer Vision*, 5(1):77–104, 1990.

[27] H. Foroosh. Pixelwise-adaptive blind optical flow assuming nonstationary statistics. *IEEE Transaction of Image Processing*, 14(2):222–230, February 2005.

[28] R.R. Fulton, S.R. Meikle, S. Eberl, J. Pfeiffer, C.J. Constable, and MJ Fulham. Correction for head movements in positron emission tomography using an optical motion-tracking system. *IEEE Transactions on Nuclear Science*, 49:116–123, 2002.

[29] S. Galic and S. Loncaric. Spatio-temporal image segmentation using optical flow and clustering. In *Proceedings of the First International Workshop on Image and Signal Processing 2000*, pages 63–68, 2000.

[30] B. Galvin, B. McCane, K. Novins, D. Mason, and S. Mills. Recovering motion fields: an evaluation of eight optical flow algorithms. In *Proceedings of the Ninth British Machine Vision Conference*, volume 1, pages 195–204, 1998.

[31] D.R. Gilland, B.A. Mair, J.E. Bowsher, and R.J. Jaszczak. Simultaneous reconstruction and motion estimation for gated cardiac ECT. *IEEE Transactions on Nuclear Science*, 49(5):2344–2349, October 2002.

[32] J.M. Gorce, D. Friboulet, and I.E. Magnin. Estimation of three-dimensional cardiac velocity fields: assessment of a differential method and application to three-dimensional CT data. *Medical Image Analysis*, 1(3):245–261, Apr 1997.

[33] N. Hata, A. Nabavi, W.M. Wells, S.K. Warfield, R. Kikinis, P.M. Black, and F.A. Jolesz. Three-dimensional optical flow method for measurement of volumetric brain deformation from intraoperative MR images. *Journal of Computer Assisted Tomography*, 24(4):531–538, 2000.

[34] B. Horn and B. Schunck. Determining optical flow. *Artificial Intelligence*, 17:185–203, 1981.

[35] G.J. Klein. Deformable Models for Volume Feature Tracking. PhD thesis, The University of California, Berkley, 1999.

[36] G.J. Klein, B.W. Reutter, and R.H. Huesman. Non-rigid summing of gated pet via optical flow. *IEEE Transactions on Nuclear Science*, 44(4):1509–1512, August 1997.

[37] T.M. Lehmann, C. Gönner, and K. Spitzer. Survey: interpolation methods in medical image processing. *IEEE Transactions on Medical Imaging*, 18(11):1049–1075, Nov 1999.

[38] T.M. Lehmann, C. Gonner, and K. Spitzer. Addendum: B-spline interpolation in medical image processing. *IEEE Transactions on Medical Imaging*, 20(7):660–665, July 2001.

[39] W. Lu and T.R. Mackie. Tomographic motion detection and correction directly in sinogram space. *Physics in Medicine and Biology*, 47(8):1267–1284, Apr 2002.

[40] B.D. Lucas and T. Kanade. An iterative image registration technique with an application to stereo vision. *Proceedings of Imaging Understanding Workshop*, pages 121–130, 1981.

[41] E. Maeland. On the comparison of interpolation methods. *IEEE Transactions on Medical Imaging*, 7(3):213–217, Sept. 1988.

[42] F. Maes, A. Collignon, D. Vandermeulen, G. Marchal, and P. Suetens. Multimodality image registration by maximization of mutual information. *IEEE Transactions on Medical Imaging*, 16(2):187–198, Apr 1997.

[43] J.B. Maintz and M.A. Viergever. A survey of medical image registration. *Medical Image Analysis*, 2(1):1–36, Mar 1998.

[44] B.A. Mair, D.R. Gilland, and J. Sun. Estimation of images and non-rigid deformations in gated emission CT. *IEEE Transactions on Medical Imaging*, 25(9):1130–1144, Sep 2006.

[45] R. Manjeshwar, X. Tao, E. Asma, and K. Thielemans. Motion compensated image reconstruction of respiratory gated PET/CT. In *IEEE International Symposium on Biomedical Imaging: From Nano to Macro*, pages 674–677, 2006.

[46] B. McCane, K. Novins, D. Crannitch, and B. Galvin. On benchmarking optical flow. *Computer Vision and Image Understanding*, 84:126–143, 2001.

[47] K. McLeish, D.L.G. Hill, D. Atkinson, J.M. Blackall, and R. Razavi. A study of the motion and deformation of the heart due to respiration. *IEEE Trans Med Imaging*, 21(9):1142–1150, Sep 2002.

[48] S.A. Nehmeh, Y.E. Erdi, C.C. Ling, K.E. Rosenzweig, H. Schoder, S.M. Larson, H.A. Macapinlac, O.D. Squire, and J.L. Humm. Effect of respiratory gating on quantifying PET images of lung cancer. *J Nucl Med*, 43(7):876–881, Jul 2002.

[49] S.A. Nehmeh, Y.E. Erdi, K.E. Rosenzweig, H. Schoder, S.M. Larson, O.D. Squire, and J.L. Humm. Reduction of respiratory motion artifacts in PET imaging of lung cancer by respiratory correlated dynamic PET: methodology and comparison with respiratory gated PET. *Journal of Nuclear Medicine*, 44(10):1644–1648, Oct 2003.

[50] D. Papathanassiou, S. Becker, R. Amir, B. Menroux, and J.C. Liehn. Respiratory motion artefact in the liver dome on FDG PET/CT: comparison of attenuation correction with CT and a caesium external source. 32(12):1422–1428, Dec 2005.

[51] G.P. Penney, J.A. Schnabel, D. Rueckert, M.A. Viergever, and W.J. Niessen. Registration-based interpolation. *IEEE Transactions on Medical Imaging*, 23(7):922–926, July 2004.

[52] Y. Picard and C.J. Thompson. Motion correction of PET images using multiple acquisition frames. *IEEE Transactions on Medical Imaging*, 16(2):137–144, Apr 1997.

[53] J.P. Pluim, J.B. Maintz, and M.A. Viergever. Image registration by maximization of combined mutual information and gradient information. *IEEE Transactions on Medical Imaging*, 19(8):809–814, Aug 2000.

[54] J.L. Prince, S.N. Gupta, and N.F. Osman. Bandpass optical flow for tagged MRI. *Medical Physics*, 27(1):108–118, Jan 2000.

[55] F. Qiao, T. Pan, J.W. Clark, and O.R. Mawlawi. A motion-incorporated reconstruction method for gated PET studies. *Physics in Medicine and Biology*, 51(15):3769–3783, Aug 2006.

[56] M. Qiu. Computing optical flow based on the mass-conserving assumption. In *Proceedings of the 15th International Conference on Pattern Recognition*, volume 3, pages 1029–1032, 2000.

[57] T.R. Reed. The computation of optical flow using the 3-D Gabor transform, *Multidimensional Systems and Signal Processing*, volume 9, pages 447–452, 2008.

[58] G. Shechter, B. Shechter, J.R. Resar, and R. Beyar. Prospective motion correction of X-ray images for coronary interventions. *IEEE Transactions on Medical Imaging*, 24(4):441–450, Apr 2005.

[59] S.M. Song and R.M. Leahy. Computation of 3-D velocity fields from 3-D cine CT images of a human heart. *IEEE Transactions on Medical Imaging*, 10:295–306, 1991.

[60] P. Thevenaz, T. Blu, and M. Unser. Interpolation revisited. *IEEE Transactions on Medical Imaging*, 19(7):739–758, Jul 2000.

[61] M. Thomas, C. Kambhamettu, C.A. Geiger, J. Hutchings, J. Richter-Menge, and M. Engram. Near-real time application of SAR-derived sea ice differential motion during APLIS ice camp 2007. In *Proceedings of the Annual Conference of the Remote Sensing & Photogrammetry Society*, 2007.

[62] D.A. Torigian, W.B. Gefter, J.D. Affuso, K. Emami, and L. Dougherty. Application of an optical flow method to inspiratory and expiratory lung MDCT to assess regional air trapping: a feasibility study. *American Journal of Roentgenology*, 188(3):W276–W280, Mar 2007.

[63] B.M.W. Tsui, W.P. Segars, and D.S. Lalush. Effects of upward creep and respiratory motion in myocardial SPECT. *IEEE Transactions on Nuclear Science*, 47:1192–1195, 2000.

[64] S.S. Vedam, V.R. Kini, P.J. Keall, V. Ramakrishnan, H. Mostafavi, and R. Mohan. Quantifying the predictability of diaphragm motion during respiration with a noninvasive external marker. *Medical Physics*, 30(4):505–513, Apr 2003.

[65] B.C. Vemuri, S. Huang, S. Sahni, C.M. Leonard, C. Mohr, R. Gilmore, and J. Fitzsimmons. An efficient motion estimator with application to medical image registration. *Medical Image Analysis*, 2(1):79–98, Mar 1998.

[66] D. Zhang and G. Lu. An edge and color oriented optical flow estimation using block matching. In *Signal Processing Proceedings*, volume 2, pages 1026–1032, 2000.

[67] Y. Zhang, D.B. Goldgof, and S. Sarkar. Towards physically-sound registration using object-specific properties for regularization. In JBA Maintz JC Gee and MW Vannier, editors, *Biomedical Image Registration*, volume 2717 of *Lecture Notes on Computer Science*, pages 358–366, Heidelberg, 2003. Springer.

[68] G.P. Zientara, P. Saiviroonporn, P.R. Morrison, M.P. Fried, S.G. Hushek, R. Kikinis, and F.A. Jolesz. MRI monitoring of laser ablation using optical flow. *Journal of Magnetic Resonance Imaging*, 8(6):1306–1318, 1998.

[69] B. Zitova and J. Flusser. Image registration methods: A survey. *Image and Vision Computing*, 21:977–1000, 2003.

Chapter 9

Combined Correction and Reconstruction Methods

Martin Benning

Institute for Computational and Applied Mathematics, University of Münster, Münster, Germany

Thomas Kösters

European Institute for Molecular Imaging, University of Münster, Münster, Germany

Frederic Lamare

Medical Research Council Clinical Sciences Centre, Imperial College London, Hammersmith Campus, London, United Kingdom

9.1	Introduction	186
9.2	Parameter identification	187
	9.2.1 Compartment modeling	187
	9.2.2 4D methods incorporating linear parameter identification	189
	9.2.3 4D methods incorporating nonlinear parameter identification	190
9.3	Combined reconstruction and motion correction	192
	9.3.1 The advantages of the list mode format	193
	9.3.2 Motion correction during an iterative reconstruction algorithm	194
	9.3.2.1 Approaches based on a rigid or affine motion model	194
	9.3.2.2 Approaches based on a non-rigid motion model	196
9.4	Combination of parameter identification and motion estimation	198
References		200

9.1 Introduction

In the beginning of PET reconstruction in the middle of the 1970s the well-known reconstruction algorithms for CT were applied since both imaging techniques rely on the efficient recovery of a function from its line integrals [40]. It turned out that the resulting images were good enough that these algorithms were use over years until new algorithms especially designed for PET were introduced [54]. As presented in Chapter 3 the filtered-backprojection type algorithms (FBP) were more and more replaced by new model-based, iterative algorithms that take into account the Poisson statistical properties [57, 22]. The new algorithms—mainly originating from the maximum-likelihood expectation-maximization (ML-EM) algorithm—show good performance even in the case of low statistic measurements where the old reconstruction algorithms may fail. Next to the statistical properties, these algorithms can easily be extended with any other corrections, such as resolution modeling, the use of different forward and backward projectors, or even higher dimensional reconstruction approaches with the use of suitable basis functions. These extensions are possible since the model-based reconstruction algorithms are not limited to the X-ray projection geometry of the FBP type algorithms. We mention that there are, as well, approaches to extend the analytical algorithms, such as the motion compensated local tomography of Katsevich [26], but usually the ML-EM-type algorithms are used for extensive modeling. In general the statistical model is used and extended for the necessary correction, whereas the resulting linear (in some cases even non-linear) equation system is solved afterwards. Due to the enormous speed increase of CPUs and porting of reconstruction algorithms to clusters or even GPUs [52] we are now able to work with extensive PET models. In this chapter we will focus on two interesting, recently growing approaches:

- Parameter estimation during 4D reconstructions using compartment models to investigate physiological parameters.

- Incorporation of motion information in the reconstruction algorithm to correct for reconstruction artifacts due to patient movement or intrinsical motion (like heart beat, breathing).

We will present both approaches in detail giving an overview of recent publications. In the end one may recognize that both approaches are multidimensional reconstructions using different types of basis functions. Hence both approaches can be combined to create an even more complex reconstruction model.

9.2 Parameter identification

In comparison to many other medical imaging techniques, positron emission tomography (PET) allows the monitoring of physiological instead of anatomical information. Physiological processes over time can often be described by intuitive mathematical models relating physiological parameters to measured PET activity. Typical applications for parameter estimation are, e.g., the determination of tumor or myocardial perfusion [51, 23, 33, 25, 1, 63, 7], or ligand-receptor-binding in the brain [19].

Although algorithms for direct parametric reconstruction go back to the mid-1980s [12], combined parameter identification and reconstruction methods did not become popular for a long period. Due to the complexity of those combined methods, these still time-demanding approaches could not be realized on the hardware available. Thus, in the early stages the parameter identification process was rather treated as a two-step procedure. First, PET reconstructions were computed for several periods in time. Subsequently, the obtained frames were preprocessed via a fitting of the desired parameters to these frames with respect to the specific model. As we will discover in the upcoming section, these models usually incorporate the temporal dependency between the frames. The drawback of the two-step procedure was the loss of temporal correlation between the PET-data sets. The combination of reconstruction and parameter identification methods has recently become a major field of interest in scientific research.

In the following we are going to present state-of-the-art compartment models used to describe physiological processes, and we are going to present recent progress in combined parameter identification/reconstruction methods.

9.2.1 Compartment modeling

Underlying models describing physiological processes such as myocardial perfusion are usually variants of so-called compartment models. Basically, compartments are homogeneous spatial regions (e.g., a voxel) for which radioactive tracer concentrations can be described as functions in time. Figure 9.1(a) shows the simplest of all compartment models, the one-compartment model. The single compartment consists of two regions, e.g., blood and tissue, with underlying radioactive concentrations. Here $h(t)$ represents the tracer concentration in blood, while $f(t)$ denotes the tracer concentration in tissue. For the specific compartment the radioactive concentration can exchange between these regions. The exchange over time can be described via the ordinary differential equation

$$\frac{d}{dt}f(t) = a\ h(t) - b\ f(t)\,, \tag{9.1}$$

for positive constants $a, b \in \mathbb{R}_{\geq 0}$. Under the natural assumption that at time $t = 0$ there is no radioactivity in the compartment (i.e., at the beginning of the scanning process the tracer has not yet been injected into the human or animal body and we have $h(0) = f(0) = 0$), Equation (9.1) can be expressed as a Laplace convolution, i.e.,

$$f(t) = (a\, h(\cdot) \otimes \exp(-b\,\cdot\,))(t) = a \int_0^t h(\tau) \exp\left(-b\,(t - \tau)\right)\, d\tau . \qquad (9.2)$$

The compartment model can easily be extended to N compartments, as presented in Figure 9.1(b). The tracer concentration in the blood pool interacts no longer with just a single region of tissue but with N different regions of tissue, and the concentration $f(t)$ is simply modeled as the sum of these interactions $f_n(t)$, i.e.,

$$f(t) = \sum_{n=1}^{N} f_n(t) ,$$

for

$$\frac{d}{dt} f_n(t) = a_n h(t) - b_n f_n(t) .$$

Equation (9.2) therefore changes to

$$f(t) = \int_0^t h(\tau) \sum_{n=1}^{N} a_n \exp\left(-b_n\,(t - \tau)\right)\, d\tau . \qquad (9.3)$$

Compartment modeling can also be used to derive more advanced models incorporating more complex physical relations among different parameters. A very simple non-linear model to quantify myocardial perfusion has been proposed in [4]. The radioactive tracer distribution in myocardial tissue, $f(t)$,

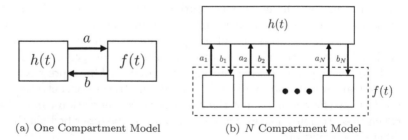

(a) One Compartment Model   (b) N Compartment Model

FIGURE 9.1: The one compartment model and its generalization to N compartments.

the tracer distribution in the left ventricular region, $h(t)$, and the myocardial perfusion, p, can be related to each other via

$$f(t) = p \int_0^t h(\tau) \exp\left(-\frac{p}{\lambda}(t - \tau)\right) \, d\tau \,, \qquad (9.4)$$

with λ denoting the so-called partition coefficient, usually set to a value between 0.9 and 1. Modifications of (9.4) have been developed, for instance [23]. Not only myocardial perfusion but also other applications involve non-linear parameter identification, as for example the measurement of regional cerebral glucose use.

However, the capabilities of compartment models are limited. Due to the restriction of investigating tracer activities in homogeneous spatial regions only, important physical effects (like motion) are completely neglected. Hence, a parameter identification of physiological parameters obtained with a model derived from compartment modeling could easily be distorted by motion artifacts. Moreover, in standard compartment modeling exchange between different spatial regions (e.g., voxels) is usually not taken into account, leading to ODE-type instead of PDE-type models.

For more information on compartment modeling we refer to [64].

9.2.2   4D reconstruction methods incorporating linear parameter identification

In [50] a method has been proposed that estimates a set of temporal basis functions and a set of coefficients for each voxel simultaneously. Two different models for the mean of the measured PET data have been used to perform an alternating minimization.

In order to combine the reconstruction process with compartment modeling, in [49] a method has been proposed that assumes every voxel to be a Laplace convolution of an abstract input curve and a set of N predefined exponential basis functions, as described in (9.3). This means, the desired image sequence is assumed to be of the type

$$f(x, t) = \sum_{n=1}^{N} a_n(x) \tilde{b}_n(t) \,, \qquad (9.5)$$

with

$$\tilde{b}_n(t) = \int_0^t h(\tau) \exp\left(-b_n (t - \tau)\right) \, d\tau \,. \qquad (9.6)$$

If we discretize the coefficients $a_n(x)$, the basis functions $\tilde{b}_n(t)$ and the input function $h(t)$ in space and time, respectively, and denote these discretizations as matrices A and B and a vector $\vec{h}$, the standard EM-algorithm can be modified to an alternating minimization scheme

$$A_{k+1} = \frac{A_k}{B_h^T K^T 1} B_h^T K^T \left(\frac{g}{K B_h A_k} \right)$$
$$\vec{h}_{k+1} = \frac{\vec{h}_k}{B_a^T K^T 1} B_a^T K^T \left(\frac{g}{K B_a \vec{h}_k} \right)$$

Here the notation B_a and B_h is indicating whether the coefficients or the input curve in the discretization of Equations (9.5) and (9.6) are kept constant during the particular EM-step. The proposed method is a combined reconstruction method; the parameters that were identified are an input curve and coefficients with respect to an exponential basis. Figure 9.2 shows $H_2^{15}O$-reconstructions of a full 4D dataset consisting of 26 3D frames, reconstructed with the standard EM-algorithm and with the modified 4D-EM-algorithm, respectively.

9.2.3   4D reconstruction methods incorporating non-linear parameter identification

Since applications such as myocardial perfusion quantification or regional cerebral glucose measurement involve the parameter identification of non-linear parameters, it is natural to combine non-linear parameter fitting and the PET reconstruction process. One option is to compute 4D EM reconstructions as described in Section 9.2.2 and to post-process these reconstructions via a non-linear parameter fitting. The obvious drawback is that the method of Section 9.2.2 is unable to capture the non-linear behavior of the particular parameters due to the linearity of the basis operators B. Hence, it seems natural to incorporate the non-linear model into the reconstruction process to enhance the parameter identification process. In [3], different variants of Equation (9.4) have been incorporated into the reconstruction process via forward-backward splitting. Results obtained from Monte-Carlo simulated synthetic $H_2^{15}O$-PET data involving parameter identifications for the parameters perfusion, arterial spillover and an arterial input function located in a left ventricular area can be seen in Figure 9.3. The data set consists of 26 temporal frames, and the spatial dimension is two- and not three-dimensional, though the whole procedure could easily be extended to full 4D. Other works on the combination of kinetic parameter estimation and PET reconstruction can be found, e.g., in [65, 66, 46, 47, 48].

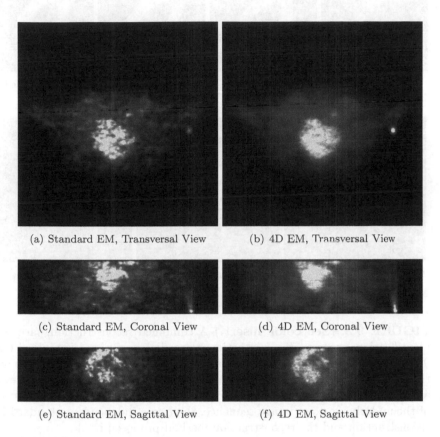

(a) Standard EM, Transversal View   (b) 4D EM, Transversal View

(c) Standard EM, Coronal View   (d) 4D EM, Coronal View

(e) Standard EM, Sagittal View   (f) 4D EM, Sagittal View

FIGURE 9.2: **(See color insert.)** Two exemplary $H_2^{15}O$ reconstructions, acquired with the standalone Siemens PET Scanner ECAT EXACT (model 921). It should be noticed that with this type of scanner image acquisition can be done only slice-by-slice. The images on the left-hand side show standard EM reconstructions of a slice of the seventh frame in transversal, coronal and sagittal view. The reconstructions have been smoothed by additional Gauss filtering between each EM iteration. The images on the right-hand side show the same views of the same slice but computed via the described 4D-EM reconstruction method. Again, after each EM iteration Gauss filtering has been applied to ensure spatial smoothness. The images on the right-hand side have a much higher signal-to-noise ratio than those derived from the frame-independent standard EM reconstructions.

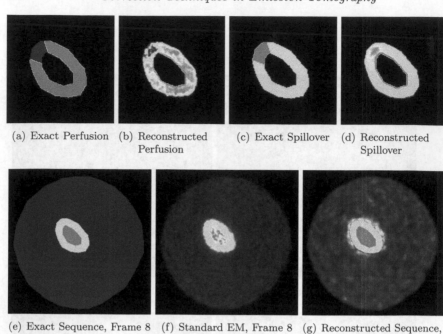

(a) Exact Perfusion   (b) Reconstructed   (c) Exact Spillover   (d) Reconstructed
                           Perfusion   Spillover

(e) Exact Sequence, Frame 8   (f) Standard EM, Frame 8   (g) Reconstructed Sequence,
                                                               Frame 8

FIGURE 9.3: (See color insert.) A combined reconstruction/parameter identification process of a synthetic myocardial perfusion example. Figure 9.3(a) and Figure 9.3(b) show the exact perfusion and the reconstruction with the method proposed in [3]. Figure 9.3(c) and Figure 9.3(d) show the exact arterial spillover and its reconstruction. Figures 9.3(e)–(g) show the 8th frame of the underlying exact image sequence, the frame-independent standard EM reconstruction and the reconstruction method proposed in [3].

9.3   Combined reconstruction and motion correction

Algorithms developed in context of Section 9.2 have also been applied to other applications where similar problems appear; e.g., Grotus et al [18] used a general 4D method to reconstruct respiratory-gated PET data. Additionally, next to the application of known methods on these new type problems, an independent theory has been investigated by several authors and will be presented in this section.

One of the parameters affecting quantitation in emission tomography (ET) imaging of the thoracic and abdominal regions is respiratory motion. Respiratory motion has been shown to reduce the accuracy of determining functional lesion volumes and associated recovered activity concentrations [41, 8] influencing positron emission tomography applications such as radiotherapy treatment planning and response to therapy monitoring respectively. Furthermore, the introduction of scanning devices combining anatomical and functional imaging has revealed various artifacts in the functional images caused by the use of the anatomical datasets for attenuation correction in combination with associated differences in the respiratory motion conditions during the acquisition of the CT and emission tomography datasets [17, 60, 13].

However, the result of multi-frame acquisitions lead to gated PET images suffering from poor signal to noise ratio since each of the frames contains only part of the counts available throughout the acquisition of a respiration average PET study [62]. A correction method applied either before or during the reconstruction process produces images with a better signal to noise ratio than a method based on image registration, especially if the reconstruction method is iterative [61, 53, 2]. Specific 4D reconstruction algorithms have to be developed to make best use of the dynamic nature of data in the reconstructed images. All these factors have therefore motivated the development of methodology that corrects for respiratory motion effects between individual gated frames in order to allow the use of the data available throughout a respiratory cycle.

9.3.1 The advantages of the list mode format

The list mode format has the advantage of containing the detection time for each line of coincidence and therefore allowing the rebinning of all the data like a dynamic acquisition. The formation of dynamic PET images usually involves a sequence of contiguous acquisitions, which are then reconstructed independently to form a final set of images that can be visualized and used to estimate the physiological parameters [21]. This approach includes the choice of the number of time bins used to rebin the acquired emission data, where a compromise has to be made between the statistical quality of images and temporal resolution: the finer the rebinning, the better the temporal resolution at the expense of the statistics used to reconstruct each image. Direct reconstruction of list mode data can avoid to make this compromise. Although with some scanners the spatial resolution is preserved with the sinogram format, in the list mode format the detection times are stored in addition to emission data generating a very high temporal resolution while retaining the full spatial resolution. One of the difficulties in the reconstruction of dynamic images from list mode data is the large number of parameters to manage. Moreover, unlike the sinogram format where data are projected into defined directions, computation time of list mode format reconstruction depends on the number of detected events, which may be of the order of a hundred million. Therefore it is important to develop rapid and convergent algorithms that can take full advantage of high spatial and temporal resolutions of the list mode format and reconstruct dynamic images with a high spatio-temporal sampling.

With the type of algorithms seeking to correct the raw data in list mode format for movement, it is possible to make a temporal rebinning as fine as desired. If, as in the case of an infrared camera detection, the movement is known in real time, it is potentially possible to correct each individual detected LOR for movement [10].

9.3.2 Motion correction during an iterative reconstruction algorithm

In this section we will describe several approaches to divide the list mode data in different gates according to the motion information and the incorporation of motion correction into the reconstruction algorithm.

9.3.2.1 Approaches based on a rigid or affine motion model

A first set of methods is based on data-driven detection and correction of movement through the adaptation of an iterative reconstruction algorithm. This methodology has previously been suggested in SPECT for cardiac applications [34] and for brain imaging [28]. These two techniques decompose in time all the acquired projections. A first image is reconstructed from one of these temporal subsets of projections. At each iteration, a subset of projections is reconstructed, corrected for the movement and the initial image then compensated for the movement is updated by considering these new projections. At each new iteration, the estimate of the image is registered with respect to the initial image, allowing the computation of the transformation to be subsequently applied to the projections. Each set of projections corresponding to the same time is corrected for movement using the computed transformations and at the end of the reconstruction, a single image is reconstructed free of any blurring effect due to movement. A similar method was also proposed in PET using the sinogram format with applications in brain imaging [6].

The second set of methods rely on the use of an external signal to detect the respiratory motion. Although few techniques use the sinogram format [56], most correction techniques in pulmonary tumor imaging in PET use the list mode format. This category of techniques is based on the realignment of all the LORs detected at the same time interval of the breathing cycle.

In a first step, it is necessary to determine the transformation that will correct the movement. This transformation can be known through the use an external detection system, or in the case of a combined PET/CT system, from the corresponding dynamic CT images synchronized with the respiration [42, 44, 35, 29, 30]. This technique is more accurate than the simple detection of the respiratory movement by an external sensor such as an infrared camera, or an impedance belt, in which case the estimate of the true movement of internal organs from an external signal recorded is not easy. Recently, in PET cardiac imaging, a technique was proposed to extract the respiratory and cardiac beating movements directly from the list mode format without the use of an external device. This method obtains good detection performance as

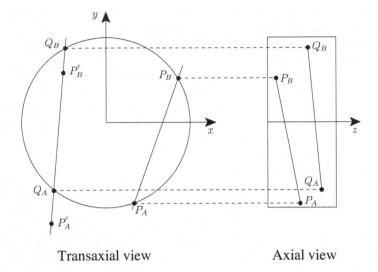

Transaxial view   Axial view

FIGURE 9.4: Representation of the affine transformation applied to a detected LOR. The scanner detector ring is approximated by a cylinder with a radius corresponding to the distance between the surface of a crystal and the PET system axis. The list mode output contains the detection coordinates of the coincidence photons, vectors $\overrightarrow{P_A}$, and $\overrightarrow{P_B}$. The affine transformation is applied on each of the individual lines of response, that is to say on the vectors $\overrightarrow{P_A}$ and $\overrightarrow{P_B}$. The corrected LOR is fully described by the two transformed detected coordinate vectors $\overrightarrow{P'_A}$ and $\overrightarrow{P'_B}$. A geometric model of the scanner is used to determine the intersections of the transformed LOR with the detector ring. If the modified LOR intersects the detector ring, the transformed detection coordinates $\overrightarrow{Q_A}$ and $\overrightarrow{Q_B}$ (intersection of the transformed LOR with the detector ring) are written in a corrected list mode file, otherwise the LOR is rejected. This method is repeated for each event of the detected list mode data sets.

long as the uptake in the heart is sufficient [11]. In PET/CT imaging, thanks to the CT images, the movement of each organ during the respiratory cycle is known. For each time interval of the respiratory cycle, the corresponding transformation is applied to each detected line of coincidence [45]; the LOR is moved back to the location it should have been detected if the patient had not moved. It is important to note that an affine (or rigid) transformation transforms a line into a line, therefore the use of such a deformation model does not modify the principle of backprojection. A geometric model of the imaging device is finaly required to determine the new detection points which correspond to the intersection between the scanner detection ring formed by the crystals and the new motion-corrected LOR [39, 29] (see Figure 9.4). This work, based on the use of an affine transformation of list mode data prior to reconstruction, has demonstrated that although this approach leads to significant improvements in lesion contrast and position in the lung fields, it is impossible to use such a model to account at the same time for respiratory motion effects in both the lung fields and organs under the diaphragm. These limitations are due to the application of a common set of affine transformation parameters across the entire emission imaging field of view, whereas the organs have independent movements due to respiration. In oncologic applications for instance, the parameters appropriate for the lungs do not significantly correct the respiratory effects for organs situated below the diaphragm, as well as the heart and the mediastinum [29].

Numerous applications have been developed for the movement correction of the head in SPECT [15] and TEP [39, 14, 16, 10]. In all these methods the movement was known through the use of an infrared camera tracking in real time the position of reflecting captors attached to the head of the patient. Moreover, as the skull is not compressible, the movement of the head is then reduced to a rotation and translation and a six-parameters transformation is therefore enough to describe the movement of the head in this type of study. An important effort has also been carried out in studying the respiratory movement in cardiac applications in TEP. The movement of the heart due to respiration is more complex than of the head. In addition to the rotation and translation, a dilation has to be considered, leading to a transformation with nine parameters. Different authors have attempted to correct for the effects of respiratory motion in cardiac emission tomography imaging through the use of either a rigid body transformation of list mode PET datasets [36] or through tracking of the center of mass in single photon emission tomography (SPECT) projections [9].

9.3.2.2   Approaches based on a non-rigid motion model

Although the application of a rigid or affine transformation in the raw data domain is feasible considering individual lines of response [36, 29], a similar approach for elastic transformation poses obvious challenges. The elastic transformation is not applied to the raw PET emission data prior to reconstruction as in the case of a rigid or affine motion model, but the iterative reconstruction algorithm has to be adapted to account for the elastic motion model. In the past, different approaches for the incorporation of transformations in the system matrix during the reconstruction process have been described [43, 24, 45, 44, 35, 30]. While the work of Qi and Huesman as well as Rahmim et al. considered only rigid body transformations, Jacobson and Fessler described the theoretical framework of incorporating non-rigid transformations in the reconstruction algorithm without evaluating the proposed methodology. In addition, Qiao et al. and Li et al. evaluated their proposed algorithm on a phantom study simulating only rigid body motion, therefore not allowing the evaluation of elastic transformations in the performance of their algorithm implementation. Finally, the patient study included in Li et al.

was at the level of the pancreas with limited respiratory motion extend of approximately 1cm and of limited non-rigid nature. Lamare et al. have proposed a methodology based on the incorporation during a list mode-based iterative reconstruction algorithm of a B-splines model allowing taking into account both the displacement of the voxels, and their shape deformation at the same time [30] (different implementations of the non-rigid transformation during the iterative reconstruction algorithm are compared in this paper, which could be of interest for some readers). They have demonstrated that the application of an elastic spatio-temporal transformation during the reconstruction process of gated PET datasets leads to significant improvements in overall image qualitative and quantitative accuracy, making use of all available data throughout a respiratory gated acquisition. In addition, their results demonstrate that the application of the spatial transformation in the raw data domain within the reconstruction leads to superior contrast (on average between 20% and 30% higher) in comparison to simply adding together already reconstructed and realigned images of the individual gated frames.

Numerous authors have previously suggested the use of 4D CT datasets to derive transformation maps subsequently used to correct for respiratory motion [35, 44, 29, 30]. Nevertheless, in a 4D PET/CT acquisition, in addition to the 4D CT images, the gated PET images are also available during a dynamic PET acquisition, and can also be used to compute the elastic motion correction parameters. Lamare et al. [31] carried out a study in order to assess the effect in terms of motion compensation in the final motion corrected image reconstructed using their motion correction integrated reconstruction [30] in combination with the transformation parameters derived from these three different dynamic image series: the 4D CT images and the gated PET images reconstructed with and without attenuation correction. The transformation parameters calculated from these three different sets of images may vary and as a result produce, after the reconstruction, images with different performances in terms of respiratory motion compensation. As can be seen in Figure 9.5, a mismatch between the dynamic CT images and the PET emission data may

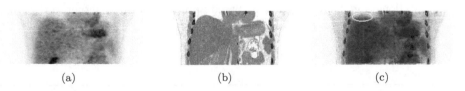

(a)   (b)   (c)

FIGURE 9.5: (See color insert.) Representation of one temporal bin of the acquisition gated with the respiration: (a) gated PET image non corrected for attenuation, (b) gated CT image, (c) overlaid image of both PET and CT gated images. A misalignment of 6/7mm between the PET and CT images can be seen at the level of the diaphragm on the overlaid image inside the yellow circle.

occur. Such a misalignment can be explained by the fact that the CT and PET acquisitions are not pursued under the same protocol; a cine CT acquisition takes usually about 30 sec corresponding to about 6 respiratory cycles, while a PET acquisition of a single bed position lasts between 4 and 12 min. As a result, it is more likely that the patients' breathing may fluctuate, both in phase and amplitude, during the CT and PET acquisitions. Accordingly, because of this difference between the CT images and PET emission data, the elastic transformation parameters calculated from the three different sets of images are not identical and when integrated in the reconstruction process; motion-compensated images are also different (see Figure 9.6). Lamare et al. [31] have demonstrated over eight clinical cases that, in the vast majority of cases, it is preferable to calculate the parameters of transformation from the PET images not corrected for attenuation rather than from attenuation-corrected PET images or CT images. Not correcting for attenuation avoids any potential mismatch that may exist between the CT images and PET emission data.

The image resolution in cardiac PET is significantly degraded due to the combination of physiological cardiac and respiratory motion [55]. With recent combined PET/CT systems, dual-gated acquisitions simultaneously synchronized with cardiac beating and respiration was proven to be feasible in both preclinical and clinical PET/CT cardiac imaging [67, 37, 27]. Moreover the motion of the heart due to respiration is not uniform across the left ventricle [38]. To account for this observation, the use of an elastic deformation algorithm has been proposed to realign individual dual-gated frames [32]. This study has demonstrated that, when trying to correct for respiratory movement only by considering either the whole cardiac cycle or the diastolic phase only, an affine or elastic motion model-based transformation applied during the reconstruction process leads to significant improvements in overall image qualitative and quantitative accuracy. However, when all the cardiac and respiratory gated frames are considered independently, the motion correction technique employed has to be able to compensate for both cardiac and respiratory motions simultaneously. In this case, the affine model shows its limitations and the elastic model based approach obtains higher performance in terms of contrast recovery and repositioning.

9.4 Combination of parameter identification and motion estimation

As explained in detail in these two sections there is a plethora of applications for parameter identification as well as for motion correction approaches in emission tomography. Both approaches were introduced as independent re-

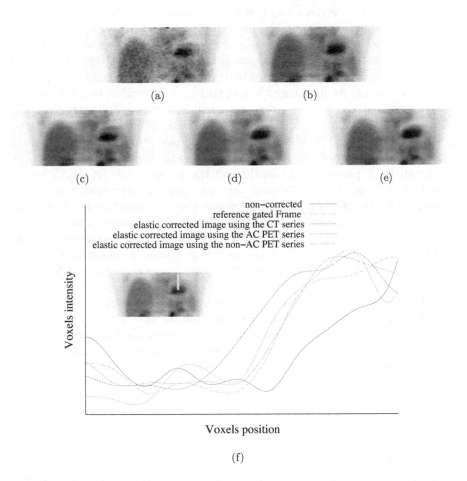

(a)   (b)

(c)   (d)   (e)

(f)

FIGURE 9.6: Comparison of (a) the gated image chosen as reference for the motion compensation reconstruction and (b) the non-corrected image, with the three motion-compensated images reconstructed using the elastic transformation parameters derived using (c) the 4D CT images, (d) the 4D PET images reconstructed with attenuation correction and (e) the 4D PET images reconstructed without attenuation correction. For the three motion-compensated images, all the PET data acquired at different time bins are elastically realigned into the specific time point corresponding to this reference-gated frame. The attenuation of the PET emission data is performed using the same 4D CT images as used for image c. As shown in graph (f), visual differences in terms of position can be clearly observed at the level of the diaphragm, below the heart.

construction/correction methods used for different applications. An obvious question arises: Are there any cases that require a combination of both approaches and is it even possible to do this?

Let us again consider the case of brain imaging. It was shown that motion affects the quantitative evaluation of dynamic PET studies whereas the amount of the quantification error depends on the magnitude of the patient's head movement. As mentioned before this kind of motion is not as complex as in case of heart imaging, but there still is the need for correction. First approaches were based on correcting each time frame independently which clearly reminds us of the first steps performed prior to the fully 4D parameter identification approaches, where all frames were reconstructed indepedently followed by a parameter estimation on the image series [20, 5]. When neglecting motion while performing the new fully 4D strategies the resulting parameter estimations may be even worse. One can imagine that motion in a single time frame leads to artifacts in all other time frames as well.

Recently, Verhaeghe et al. [58] proposed a 4D PET reconstruction algorithm for parameter identification including motion compensation. For this new approach, the 4D framework of the same group was extended for motion correction using PET data supersets [59]. It was shown that motion correction in 4D applications is necessary to incorporate into the reconstruction process on the one hand and that it can be computed on the other hand.

Although 5D reconstruction methods are rarely used in present applications they will be in focus in the next years following the trend to include as much information as possible in the reconstruction algorithms.

References

[1] J. Y. Ahn, D. S. Lee, J. S. Lee, S. Kim, G. J. Cheon, J. S. Yeo, S. Shin, J. Chung, and M. C. Lee. Quantification of regional myocardial blood flow using dynamic H215O PET and factor analysis. *Journal of Nuclear Medicine*, 42(5):782–787, 2001.

[2] E. Asma, R. Manjeshwar, and K. Thielemans. Theoretical comparison of motion correction techniques for PET image reconstruction. *IEEE Nuclear Science Symposium Conference Record*, 3:1762–1767, 2006.

[3] M. Benning, T. Kösters, F. Wübbeling, K. Schäfers, and M. Burger. A nonlinear variational method for improved quantification of myocardial blood flow using dynamic $H_2^{15}O$ PET. *IEEE Nuclear Science Symposium Conference Record*, November 2008.

[4] S. R. Bergmann, K. A. Fox, and A. L. Rand. Quantification of regional myocardial blood flow in vivo with $H_2\ ^{15}O$. *Circulation*, 70:724–733, 1984.

[5] P. M Bloomfield, T. J. Spinks, J. Reed, L. Schnorr, A. M. Westrip, L. Livieratos, R. Fulton, and T. Jones. The design and implementation of a motion correction scheme for neurological PET. *Phys. Med. Biol.*, 48:959–978, 2003.

[6] M. Blume, M. Rafecas, S. Ziegler, and N. Navab. Combined motion compensation and reconstruction for PET. *IEEE Nuclear Science Symposium Conference Record*, pages 5485–5487, 2008.

[7] R. Boellaard, P. Knaapen, A. Rijbroek, G. J. J. Luurtsema, and A. A. Lammertsma. Evaluation of basis function and linear least squares methods for generating parametric blood flow images using ^{15}O-water and positron emission tomography. *Molecular Imaging and Biology*, 7:273–285, 2005.

[8] L. Boucher, S. Rodrigue, R. Lecomte, and F. Benard. Respiratory gating for 3-dimensional PET of the thorax: feasibility and initial results. *J Nucl Med*, 45(2):214–219, 2004.

[9] P. P. Bruyant, M. A King, and P. H. Pretorius. Correction of the respiratory motion of the heart by tracking of the center of mass of thresholded projections: a simulation study using the dynamic MCAT phantom. *IEEE Transactions on Nuclear Science*, 49:2159–2166, 2003.

[10] P. Buhler, U. Just, E. Will, J. Kotzerke, and J. van den Hoff. An accurate method for correction of head movement in PET. *IEEE Transactions on Medical Imaging*, 23(9):1176–1185, 2004.

[11] F. Buther, M. Dawood, L. Stegger, F. Wubbeling, M. Schafers, O. Schober, and K. P. Schafers. List mode-driven cardiac and respiratory gating in PET. *Journal of Nuclear Medicine*, 50(5):674–81, 2009.

[12] R. E. Carson and K. Lange. A statistical model for positron emission tomography. *Journal of the American Statistical Association*, 80(389):20–22, 1985.

[13] Y. E. Erdi, S. A. Nehmeh, T. Pan, A. Pevsner, K. E. Rosenzweig, G. Mageras, E. D. Yorke, H. Schoder, W. Hsiao, O. D. Squire, P. Vernon, J. B. Ashman, H. Mostafavi, S. M. Larson, and J. L. Humm. The CT motion quantitation of lung lesions and its impact on PET-measured SUVs. *Journal of Nuclear Medicine*, 45(8):1287–1292, 2004.

[14] R. R. Fulton, B. F. Hutton, M. Braun, B. Ardekani, and R. Larkin. Use of 3D reconstruction to correct for patient motion in SPECT. *Physics in Medicine and Biology*, 39(3):563–574, 1994.

[15] R. R. Fulton, S. Eberl, S. R. Meikle, B. F. Hutton, and M. Braun. A practical 3D tomographic method for correcting patient head motion in clinical SPECT. *IEEE Transactions on Nuclear Science*, 46(3):667–672, 1999.

[16] R. R. Fulton, S. R. Meikle, S. Eberl, J. Pfeiffer, C. Constable, and M. J. Fulham. Correction for head movements in positron emission tomography using an optical motion tracking system. *IEEE Medical Imaging Conference Records*, 3:1758–1762, 2000.

[17] G. W. Goerres, E. Kamel, T. N. Heidelberg, M. R. Schwitter, C. Burger, and G. K. von Schulthess. PET-CT image co-registration in the thorax: influence of respiration. *European Journal of Nuclear Medicine and Molecular Imaging*, 29(3):351–360, 2002.

[18] N. Grotus, A. J. Reader, S. Stute, J. C. Rosenwald, P. Giraud, and I. Buvat. Fully 4D list-mode reconstruction applied to respiratory-gated PET scans. *Physics in Medicine and Biology*, 54:1705–1721, 2009.

[19] R. N. Gunn, A. A. Lammertsma, S. P. Hume, and V. J. Cunningham. Parametric imaging of ligand-receptor binding in PET using a simplified reference region model. *Neuroimage*, 6:279–287, 1997.

[20] H. Herzog, L. Tellmann, R. Fulton, I. Stangier, E. R. Kops, K. Bente, C. Boy, R. Hurlemann, and U. Pietrzyk. Motion artifact reduction on parametric PET images of neuroreceptor binding. *Journal of Nuclear Medicine*, 46(6):1059–1065, 2005.

[21] S. Huang and M. Phelps. Principles of tracer kinetic modeling in positron emission tomography and autoradiography. In *Positron Emission Tomography and Autoradiography: Principles and Applications for the Brain and Heart*, eds. Phelps, Mazziotta, and Schelbert, 287–346, New York: Raven Press, 1986.

[22] H. M. Hudson and R. S. Larkin. Accelerated image reconstruction using ordered subsets of projection data. *IEEE Transactions on Medical Imaging*, 1994.

[23] H. Iida, I. Kanno, A. Takahashi, S. Miura, M. Murakami, K. Takahashi, Y. Ono, F. Shishido, A. Inugami, and N. Tomura. Measurement of absolute myocardial blood flow with H215O and dynamic positron-emission tomography. Strategy for quantification in relation to the partial-volume effect [published erratum appears in *Circulation* 1988 Oct;78(4):1078]. *Circulation*, 78(1):104–115, 1988.

[24] M. W. Jacobson and J. A. Fessler. Joint estimation of image and deformation parameters in motion-corrected PET. *IEEE Nuclear Science Symposium Conference Record*, 5:3290–3294, 2003.

[25] C. Katoh, K. Morita, T. Shiga, N. Kubo, K. Nakada, and N. Tamaki. Improvement of algorithm for quantification of regional myocardial blood flow using 15O-water with PET. *Journal of Nuclear Medicine*, 45(11):1908–1916, 2004.

[26] A. Katsevich. Motion compensated local tomography. *Inverse Problems*, 24:1–21, 2008.

[27] T. Kokki, M. Teras, H.T. Sipila, T. Noponen, and J. Knuuti. Dual gating method for eliminating motion related inaccuracies in cardiac PET. *Nuclear Science Symposium Conference Records*, 5:3871–3875, 2007.

[28] A. Z. Kyme, B. F. Hutton, R. L. Hatton, D. W. Skerrett, and L. R. Barnden. Practical aspects of a data-driven motion correction approach for brain SPECT. *IEEE Transactions on Medical Imaging*, 22(6):722–729, 2003.

[29] F. Lamare, T. Cresson, J. Savean, C. Cheze-Le Rest, A. J. Reader, and D. Visvikis. Respiratory motion correction for PET oncology applications using affine transformation of list mode data. *Physics in Medicine and Biology*, 52(1):121–140, 2007.

[30] F. Lamare, M. Ledesma-Carbayo, T. Cresson, G. Kontaxakis, A. Santos, C. Cheze-Le Rest, A. J. Reader, and D. Visvikis. List mode based image reconstruction for respiratory motion correction in PET using non-rigid body transformations. *Physics in Medicine and Biology*, 52:5187–5204, 2007.

[31] F. Lamare, M. Ledesma-Carbayo, A. J Reader, V. Bettinardi, H. Bammer, O. Malwawi, G. Kontaxakis, A. Santos, C. Cheze-Le Rest, and D. Visvikis. Performance of an image reconstruction algorithm incorporating continuous B-spline functions in the system matrix for respiratory motion correction in 4D PET/CT. *Journal of Nuclear Medicine*, 48(2):197P, 2007.

[32] F. Lamare, O. Rimoldi, M. Teras, T. Kokki, J. Knuuti, D. Visvikis, and P. G. Camici. Pre-clinical evaluation of a respiratory motion correction technique in dual gated cardiac imaging. *Journal of Nuclear Medicine*, 50:1470, 2009.

[33] J. S. Lee, D. S. Lee, J. Y. Ahn, G. J. Cheon, S. Kim, J. S. Yeo, K. Seo, K. S. Park, J. Chung, and M. C. Lee. Blind separation of cardiac components and extraction of input function from H215O dynamic myocardial PET using independent component analysis. *Journal of Nuclear Medicine*, 42(6):938–943, 2001.

[34] T. S. Lee, P. W. Segars, and B. W. Tsui. Application of 4D MAP-RBI-EM with space time Gibbs priors to gated myocardial SPECT. *Journal of Nuclear Medicine*, 46(162P), 2005.

[35] T. Li, B. Thorndyke, E. Schreibmann, Y. Yang, and L. Xing. Model-based image reconstruction for four-dimensional PET. *Medical Physics*, 33(5):1288–98, 2006.

[36] L. Livieratos, L. Stegger, P. M. Bloomfield, K. Schafers, D. L. Bailey, and P. G. Camici. Rigid-body transformation of list-mode projection data for respiratory motion correction in cardiac PET. *Physics in Medicine and Biology*, 50(14):3313–3322, 2005.

[37] A Martinez-Moller, D. Zikic, R.M. Botnar, R.A. Bundschuh, W. Howe, S. Ziegler, N. Navab, M. Schwaiger, and Nekolla S.G. Dual cardiac-respiratory gated PET: implementation and results from a feasibility study. *European Journal of Nuclear Medicine and Molecular Imaging*, 34:1447–1454, 2007.

[38] K. McLeish, D. L. Hill, D. Atkinson, J. M. Blackall, and R. Razavi. A study of the motion and deformation of the heart due to respiration. *IEEE Transactions on Medical Imaging*, 21(9):1142–50, 2002.

[39] M. Menke, M.S. Atkins, and K.R. Buckley. Compensation methods for head motion detected during PET imaging. *IEEE Transactions on Nuclear Science*, 43(1):310–316, 1996.

[40] F. Natterer. The mathematics of computerized tomography. *Classics in Applied Mathematics*. 32. Philadelphia, PA: SIAM, 2001.

[41] S. A. Nehmeh, Y. E. Erdi, C.C. Ling, K. E. Rosenzweig, O. Squire, L.E. Braban, E. Ford, K. Sidhu, G.S. Mageras, S. M. Larson, and J. L. Humm. Effect of respiratory gating on reducing lung motion artefacts in PET imaging of lung cancer. *Medical Physics*, 29:366–371, 2002.

[42] S. A. Nehmeh, Y. E. Erdi, T. Pan, E. Yorke, G. S. Mageras, K. E. Rosenzweig, H. Schoder, H. Mostafavi, O. Squire, A. Pevsner, S. M. Larson, and J. L. Humm. Quantitation of respiratory motion during 4D-PET/CT acquisition. *Medical Physics*, 31(6):1333–1338, 2004.

[43] J. Qi and R.H. Huesman. List mode reconstruction for PET with motion compensation: a simulation study. *Proceedings IEEE International Symposium on Biomedical Imaging*, 413–416, 2002.

[44] F. Qiao, T. Pan, J. W. Clark, Jr, and O. R. Mawlawi. A motion-incorporated reconstruction method for gated PET studies. *Physics in Medicine and Biology*, 51(15):3769–83, 2006.

[45] A. Rahmim, P. Bloomfield, S. Houle, M. Lenox, C. Michel, K. R. Buckley, T. J. Ruth, and V. Sossi. Motion compensation in histogram-mode and list-mode reconstructions: beyond the event-driven approach. *IEEE Transactions on Nuclear Science*, 51(5):2588–2596, 2004.

[46] A. Rahmim, Y. Zhou, and D. F. Wong. Direct 4D parametric image reconstruction with plasma input and reference tissue models in reversible binding imaging. *IEEE Nuclear Science Symposium Conference Record*, 2516–2522, 2009.

[47] Arman Rahmim and Jing Tang. 4D PET: beyond conventional dynamic PET imaging. *Iran Journal of Nuclear Medicine*, 16(1):1–13, 2008.

[48] Arman Rahmim, Jing Tang, and Habib Zaidi. Four-dimensional (4D) image reconstruction strategies in dynamic PET: Beyond conventional independent frame reconstruction. *Medical Physics*, 36:3654–3670, 2009.

[49] A. J. Reader, J.C. Matthews, F. C. Sureau, C. Comtat, R. Trébossen, and I. Buvat. Fully 4D image reconstruction by estimation of an input function and spectral coefficients. *IEEE Nuclear Science Symposium Conference Record*, 3260–3267, 2007.

[50] A. J. Reader, F. C. Sureau, C. Comtat, R. Trébossen, and I. Buvat. Joint estimation of dynamic PET images and temporal basis functions using fully 4D ML-EM. *Physics in Medicine and Biology*, 51(21):5455–5474, 2006.

[51] K. P. Schäfers, T. J. Spinks, P. G. Camici, P. M. Bloomfield, C. G. Rhodes, M. P. Law, C. S. R. Baker, and O. Rimoldi. Absolute quantification of myocardial blood flow with H2150 and 3-dimensional PET: an experimental validation. *Journal of Nuclear Medicine*, 43(8):1031–1040, 2002.

[52] M. Schellmann, S. Gorlatch, D. Meilnder, T. Kösters, K. Schäfers, F. Wübbeling, and M. Burger. Parallel medical image reconstruction: from graphics processors to grids. *Pact09*, 2009.

[53] H. Schoder, Y. E. Erdi, K. Chao, M. Gonen, S. M. Larson, and H. W. Yeung. Clinical implications of different image reconstruction parameters for interpretation of whole-body PET studies in cancer patients. *Journal of Nuclear Medicine*, 45(4):559–566, 2004.

[54] L.A. Shepp and Y. Vardi. Maximum likelihood reconstruction for emission tomography. *IEEE Transactions on Medical Imaging*, 1(2):113122, 1982.

[55] M. M. Ter-Pogossian, S. R. Bergmann, and B. E. Sobel. Influence of cardiac and respiratory motion on tomographic reconstructions of the heart: implications for quantitative nuclear cardiology. *Journal of Computer Assisted Tomography*, 6(6):1148–55, 1982.

[56] K. Thielemans, S. Mustafovic, and L. Schnorr. Image reconstruction of motion corrected sinograms. *IEEE Medical Imaging Conference Record*, 2401–2406, 2003.

[57] Y. Vardi, L. A. Shepp, and L. Kaufman. A statistical model for positron emission tomography. *Journal of the American Statistical Association*, 80:8–20, 1985.

[58] J. Verhaeghe, P. Gravel, R. Mio, R. Fukasawa, P. Rosa-Neto, J.-P. Soucy, C. J. Thompson, and A. J. Reader. Motion-compensated fully 4D PET reconstruction using PET data supersets. In *Proc. IEEE Nuclear Science Symposium and Medical Imaging Conference*, 2009.

[59] J Verhaeghe and Andrew J Reader. PET projection data supersets for reconstruction with acquisition motion. In *Proc. IEEE Nuclear Science Symposium and Medical Imaging Conference*, 2009.

[60] D. Visvikis, O. Barret, T.D. Fryer, F. Lamare, A. Turzo, Y. Bizais, and C. Cheze-Le Rest. Evaluation of respiratory motion effects in comparison with other parameters affecting PET image quality. *IEEE Nuclear Science Symposium Conference Record*, 6:3668–3672, 2004.

[61] D. Visvikis, C. Cheze-Le Rest, D. C. Costa, J. Bomanji, S. Gacinovic, and P. J. Ell. Influence of OSEM and segmented attenuation correction in the calculation of standardised uptake values for [18F]FDG PET. *European Journal of Nuclear Medicine*, 28(9):1326–1335, 2001.

[62] D. Visvikis, F. Lamare, A. Turzo, O. Barrett, T. Fryer, Y. Bizais, and C. Cheze-Le Rest. Efficiency of respiratory gating for motion correction in PET. *Journal of Nuclear Medicine*, 46(2):163P, 2005.

[63] H. Watabe, H. Jino, N. Kawachi, N. Teramoto, T. Hayashi, Y. Ohta, and H. Iida. Parametric imaging of myocardial blood flow with 15O-water and PET using the basis function method. *Journal of Nuclear Medicine*, 46(7):1219–1224, 2005.

[64] M. N. Wernick and J. N. Aarsvold. *Emission Tomography: The Fundamentals of PET and SPECT*. Elsevier Academic Press, 2004.

[65] J. Yan, B. Planeta-Wilson, and R. E. Carson. Direct 4D list mode parametric reconstruction for PET with a novel EM algorithm. *IEEE Nuclear Science Symposium Conference Record*, 3625–3628, 2008.

[66] J. Yan, B. Planeta-Wilson, J. D. Gallezot, and R. E. Carson. Initial evaluation of direct 4D parametric reconstruction with human PET data. *IEEE Nuclear Science Symposium Conference Record*, 2503–2506, 2009.

[67] Y. Yang, S. Rendig, S. Siegel, D. F. Newport, and S. R. Cherry. Cardiac PET imaging in mice with simultaneous cardiac and respiratory gating. *Physics in Medicine and Biology*, 50(13):2979–89, 2005.

Part III

Recent Developments

Chapter 10

Introduction Hybrid Tomographic Imaging

Hartwig Newiger

Siemens Healthcare, Molecular Imaging, Erlangen, Germany

10.1   Introduction 209
10.2   Combining PET and SPECT 209
10.3   The combination with MR 211
10.4   Combining ultrasound with PET and SPECT 213
References 215

10.1   Introduction

When the December 4, 2000 issue of *Time* magazine listed PET/CT as one of the key innovations in 2000, it was not obvious that within about four years nearly 100% of all newly installed PET systems would be PET/CT devices. But PET/CT, developed by Townsend and his team, has changed the world of PET forever [1]. The combination of highly specific metabolic imaging with fine spatial and temporal resolution of the anatomic information led to enormous improvements in the field of PET imaging (and PET/CT), providing diagnostic information far beyond the individual results of either modality (Figure 10.1).

From PET/CT it was only a small step to combine SPECT devices with diagnostic CT (Figure 10.2) which was introduced commercially in 2004 to demonstrate its clinical efficacy [7].

Since then, hybrid imaging has become a standard tool in nuclear medicine, combining functional information with the morphological data sets of diagnostic CT.

Some diagnostic questions require the correlation of PET and SPECT studies with data from other (anatomical) imaging devices beyond CT. Examples are the combination with MR or ultrasound (US) or the comparison of PET and SPECT directly [9].

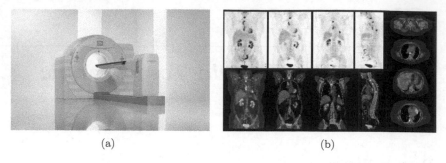

(a)   (b)

FIGURE 10.1: (See color insert.) PET/CT. (a) Siemens Biograph® mCT. (b) 5-min ultraHD PET study of an obese patient with lung CA (data courtesy of University of Tennessee, Knoxville, TN, USA).

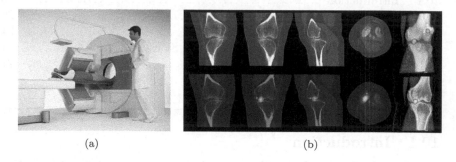

(a)   (b)

FIGURE 10.2: (See color insert.) SPECT/CT. (a) Siemens Symbia® T16 Truepoint™SPECT/CT. (b) SPECT/CT delineation of subchondral cyst in left knee joint (data courtesy of PRP Cumberland Diagnostic Imaging, New South Wales, Australia).

10.2   Combining PET and SPECT

Today, the direct combination of SPECT and PET is available only for animal imaging with devices like the Inveon® [4]. Such a combination (Figure 10.3) allows optimization of the tracers used to visualize selected biochemical processes. While PET tracers are usually labeled with relatively short-lived positron emitters, there are several single photon emitters with longer half-life that enable the characterization of metabolic pathways with longer time factors.

Such systems can also be combined with a CT device to overlay the PET and/or SPECT information with the anatomical structure of the specimen studied. Clinical studies would have to be conducted on individual

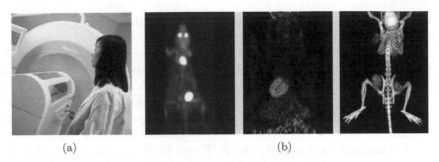

(a) (b)

FIGURE 10.3: (See color insert.) PET-SPECT-CT. (a) Siemens Inveon®
MultiModality. (b) Preclinical studies (courtesy of University of Wisconsin,
Madison, WI and Eberhard Karls University Tuebingen, Tuebingen, Germany).

SPECT(/CT) and PET(/CT) systems and image fusion would then have to
be done during the processing step retrospectively.

10.3   The combination with MR

In the 1980s, magnetic resonance imaging (MRI) was introduced to clinical
practice. It quickly became a standard tool in the diagnosis of soft-tissue
abnormalities due to its high contrast and the ability to acquire easily oblique
slices. The absence of ionizing radiation also helped the acceptance of this new
modality.

Magnetic resonance uses the precession of particles with a spin and magnetic momentum in a magnetic field. A strong static magnetic field aligns these
particles. When applying RF pulses at the resonance frequency, the particles
will be pushed out of their stable condition absorbing the electromagnetic
energy, but will then return back to the baseline by emitting the absorbed
electromagnetic energy. The relaxation time and the amplitude of this signal is related to the chemical condition of the particle and its concentration.
Images are reconstructed based on either the longitudinal relaxation time T1
(related to magnetic moments of surrounding nuclei) or the transversal relaxation time T2 (determined by the frequency of the collisions of the molecules)
or a combination of both. The most common nucleus used for MR-imaging is
hydrogen, but other nuclei are used as well.

While general MR imaging provides superb images with high soft-tissue
contrast, more sophisticated pulse sequences and imaging techniques allow,
for example, flow and perfusion images, determination of spatial distribu-

tion of certain molecules (MR spectroscopy), functional imaging (fMRI), and diffusion-tensor tracking [6].

The combination of MR and PET is very attractive. High soft-tissue contrast (MRI) combined with specific PET tracers may lead to completely new and effective work flows and higher diagnostic accuracy. High sensitivity and specificity are expected. In addition, acquisition times of PET and MRI are in the same order of magnitude and thus this coupling suggests several synergetic advantages. But the integration is also technically very challenging.

A simultaneous acquisition of PET and MRI images requires full PET-hardware compatibility to MR. This means, for example, that the standard PET detector (i.e., scintillation crystal coupled to a photomultiplier tube (PMT)) has to be replaced by a combination of the scintillator and a photo avalanche diode or SiPM, i.e., a detector that is significantly less sensitive to magnetic fields. Another challenge is the fact that quantitative PET imaging requires that the PET data are corrected for attenuation. While the CT information can easily be scaled to correct for attenuation and scatter [10], this is not quite that simple for MRI data sets. Unfortunately, the standard MRI acquisition shows only very little bone contrast, one of the sources of PET attenuation. In addition, the soft-tissue information of the MRI image is not correlated to the attenuation coefficient necessary for PET attenuation correction [2, 3]. Either specific MR pulse sequences have to be implemented to enable *CT*—such as datasets which could directly be used for attenuation correction or theoretical patient models (e.g., based on an atlas) have to be implemented. Special care has to be taken when magnetic and paramagnetic material is used for implants. Some of these implants may exclude the patient from an MR-PET scan altogether; others will lead to significant artifacts in the MR data which do require specific software tools to align the MR information for PET attenuation correction and to guarantee a perfect match of the PET and the MR information.

On the other hand, a simultaneous acquisition of PET and MR may allow for even more sophisticated data corrections. One possibility would be an online motion correction of the PET information due to the fact that fast MR sequences are a very sensitive tool to detect and track organ and patient movement.

First prototypes of such a combined MR-PET system (Figures 10.4 and 10.5) have demonstrated the performance of this type of hybrid imaging device and are being used to explore the clinical possibilities. While these initial systems are limited to brain applications, they nevertheless allow the extrapolation to whole-body designs [8, 5].

At RSNA 2010 the first fully integrated whole-body MR-PET system has been introduced combining state of the art 3T MR technology with high performance PET technology for simultaneous MR-PET whole-body applications. A prototype system is demonstrating first results at the Technical University of Munich, University Hospital rechts der Isar, Department of Nuclear Medicine (Figure 10.6).

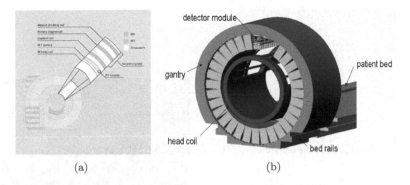

(a)   (b)

FIGURE 10.4: (See color insert.) MR-PET: (a) detector layout of Siemens BrainPET™(prototype) and (b) design study of whole-body MR-PET.

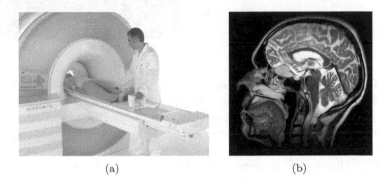

(a)   (b)

FIGURE 10.5: (See color insert.) MR-PET: Prototype. (a) Siemens prototype BrainPET™(Works in Progress, The product is under development and is not commercially available. (b) MR-PET study (courtesy of University Tübingen, Germany).

Combining MR with SPECT will add additional challenges. SPECT systems typically require a rotating/moving collimator (with or without a moving scintillator) to allow the acquisition of all the projections necessary to reconstruct the tomographic images. A solution for such an MR-compliant collimator needs still to be found. For MR-SPECT a side-by-side design may be the only feasible one, of course prohibiting a simultaneous acquisition of SPECT and MR data sets.

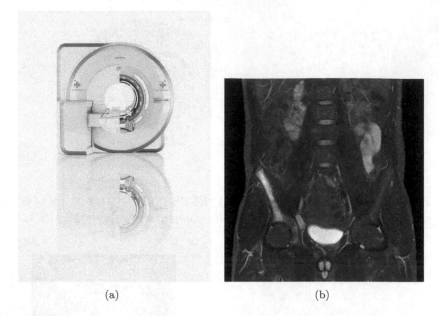

(a)   (b)

FIGURE 10.6: (See color insert.) Fully integrated whole-body MR-PET. (a) Siemens Biograph™mMR; the Biograph™mMR system requires 510(k) review by the FDA and is not commercially available. Due to regulatory reasons its future availability in any country cannot be guaranteed. Please contact your local Siemens organization for further details. (b) Simultaneously acquired MR-PET showing bone marrow imaging in case of cancer.

10.4   Combining ultrasound with PET and SPECT

The real-time visualization of soft-tissue structures, movements and flow makes ultrasound (US) a very flexible tool. Recent developments have made diagnosis less user dependent and allow three-dimensional images of structure and function of the volume to be diagnosed in real-time. A combination with PET and/or SPECT with us imaging could yield additional data to better quantify the nuclear medicine information. The us acquisition, though, requires a direct contact of the probe to the skin. Only then can the us beam be transmitted through the tissue and the probe detect the reflections from the varying tissue and structure boundaries. These constraints most likely will prohibit a simultaneous acquisition of PET or SPECT with us, especially as the US probe will be hard to design to only use materials low in attenuation for the photons used for PET and SPECT imaging.

Most likely the information will have to be fused using information of the

3D US volume acquired individually and registered retrospectively to the PET and/or SPECT volume.

References

[1] T. Beyer, D.W. Townsend, T. Brun, P.E. Kinahan, M. Charron, R. Roddy, J. Jerin, J. Young, L. Byars, and R. Nutt. A combined PET/CT scanner for clinical oncology. *Journal of Nuclear Medicine*, 41(8):1369, 2000.

[2] T. Beyer, M. Weigert, H.H. Quick, U. Pietrzyk, F. Vogt, C. Palm, G. Antoch, S.P. Müller, and A. Bockisch. MR-based attenuation correction for torso-PET/MR imaging: pitfalls in mapping MR to CT data. *European Journal of Nuclear Medicine and Molecular Imaging*, 35(6):1142–1146, 2008.

[3] C. Catana, A. van der Kouwe, T. Benner, C.J. Michel, M. Hamm, M. Fenchel, B. Fischl, B. Rosen, M. Schmand, and A.G. Sorensen. Toward implementing an MRI-Based PET attenuation-correction method for neurologic studies on the MR-PET brain prototype. *Journal of Nuclear Medicine*, 51(9):1431, 2010.

[4] S.S. Gleason, D.W. Austin, R.S. Beach, R. Nutt, M.J. Paulus, and S. Yan. A new highly versatile multimodality small animal imaging platform. In *IEEE Nuclear Science Symposium Conference Record, 2006*, volume 4, 2006.

[5] M.S. Judenhofer, H.F. Wehrl, D.F. Newport, C. Catana, S.B. Siegel, M. Becker, A. Thielscher, M. Kneilling, M.P. Lichy, M. Eichner, W.D. Heiss, and C.D. Claussen. Simultaneous PET-MRI: a new approach for functional and morphological imaging. *Nature Medicine*, 14(4):459–465, 2008.

[6] A. Oppelt. *Imaging systems for medical diagnostics: fundamentals, technical solutions and applications for systems applying ionization radiation, nuclear magnetic resonance and ultrasound.* Publicis Corporate Pub., 2005.

[7] W. Romer, A. Nomayr, M. Uder, W. Bautz, and T. Kuwert. SPECT-guided CT for evaluating foci of increased bone metabolism classified as indeterminate on SPECT in cancer patients. *Journal of Nuclear Medicine*, 47(7):1102, 2006.

[8] H.P.W. Schlemmer, B.J. Pichler, M. Schmand, Z. Burbar, C. Michel, R. Ladebeck, K. Jattke, D. Townsend, C. Nahmias, P.K. Jacob, W.D.

Heiss, and C.D. Claussen. Simultaneous MR/PET imaging of the human brain: feasibility study. *Radiology*, 248(3):1028, 2008.

[9] D.W. Townsend. Multimodality imaging of structure and function. *Physics in Medicine and Biology*, 53:R1, 2008.

[10] C.C. Watson, M.E. Casey, C. Michel, B. Bendriem. Advances in scatter correction for 3D PET/CT. In *2004 IEEE Nuclear Science Symposium Conference Record*, pages 3008–3012, 2004.

Chapter 11

MR-based Attenuation Correction for PET/MR

Matthias Hofmann
Max Planck Institute for Biological Cybernetics, Tübingen, Germany
Laboratory for Preclinical Imaging and Imaging Technology, University of Tübingen, Tübingen, Germany
Department of Engineering Science, University of Oxford, Oxford, United Kingdom

Bernd Pichler
Laboratory for Preclinical Imaging and Imaging Technology, University of Tübingen, Tübingen, Germany

Thomas Beyer
cmi-experts GmbH, Zürich, Switzerland

11.1	Introduction	218
11.2	MR-AC for brain applications	220
	11.2.1 Segmentation approaches	220
	11.2.2 Atlas approaches	221
11.3	Methods for torso imaging	224
11.4	Discussion	229
	11.4.1 The presence of bone	230
	11.4.2 MR imaging with ultrashort echo time (UTE)	231
	11.4.3 Required PET accuracy	232
	11.4.4 Validation of MR-AC methods	232
	11.4.5 Truncated field-of-view	232
	11.4.6 MR coils and positioning aids	233
	11.4.7 User intervention	233
	11.4.8 Potential benefits of MR-AC	234
	11.4.9 Additional potential benefits of simultaneous PET/MR acquisition	234
11.5	Conclusion	234
References		235

11.1   Introduction

Current concepts of combined PET/MR tomographs do not allow for separate CT-like transmission sources. Therefore, corresponding PET attenuation coefficients must be calculated from the available MR images. MR-based attenuation correction (MR-AC) is far more challenging than the well-established algorithms for CT-based attenuation correction (CT-AC) since MR image voxel values correlate with the hydrogen nuclei density in tissues and tissue relaxation properties rather than with electron density-related mass attenuation coefficients on CT (Figure 11.1). Therefore, a direct transformation from available MR images to CT-like attenuation values is challenging [3].

Although pre-clinical PET/MR prototype systems [37] have been around since the mid 1990s MR-AC is still a work in progress. While early pre-clinical PET/MR design concepts did not include means for AC [17], [5] a relatively simple 2-class attenuation correction scheme was suggested for the first clinical prototype [36]. The lack of viable MR-AC methods today can be explained by the fact that attenuation is less critical in small animals compared to pediatric and adult patients, and, therefore, the issue of AC was of minor importance in pre-clinical PET and PET/MR.

In this chapter we review methods to derive PET attenuation maps from available MR images in clinical PET/MR imaging scenarios. We will discuss potential pitfalls of MR-AC as well as a number of possible advantages of MR-AC that in certain cases could render it beneficial over CT-AC.

Table 11.1 summarizes the main approaches to MR-AC for clinical imaging scenarios. Interestingly, several studies of MR-AC appeared when such com-

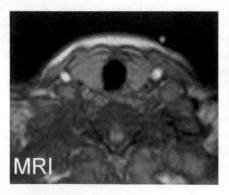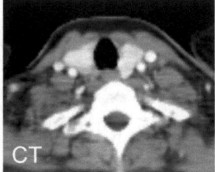

FIGURE 11.1: Axial MR and CT images of the neck of the same patient showing the differences in appearance of bone, air (trachea) and soft tissue. The inability to segment bone structures based only on their MR intensity renders simple scaling approaches to MR-AC impossible.

TABLE 11.1: Overview of studies on MR-AC. (All studies compare PET with standard PET attenuation correction and MR-AC. Note, different studies used different reconstruction and correction techniques and, therefore, the results may not be directly comparable.)

Area	Author, Year	Methodology	Results	Ref
Head	Rougetet, 1994	Phantom studies + 1 patient; 0.5T MR with T1w; 3-class MR-segmentation and mu-assignment (skin: 0.095 cm-1, brain: 0.095 cm-1, bone: 0.151 cm-1)	Max. difference between PET_{TXAC} and PET_{MRAC} was 11% and 12% in phantoms and patient, respectively. Future: Investigate co-registration of emission-only and MRI; improve MR segmentation of the lower skull	[6]
Head	Zaidi, 2003	10 patients; 1.5 T MRI with T1w; 5-class fuzzy-information segmentation of MRI and mu-assignment (air: 0 cm-1, brain tissue: 0.0993 cm-1, skull: 0.143 cm-1, nasal sinuses: 0.0553 cm-1, scalp: not considered)	R^2= 0.91 for ROIs in PET_{TXAC} and PET_{MRAC}. Results indicate an improvement in PET image quality following MR-AC. Future: Integrate MR information into a unified statistical PET reconstruction method	[7]
Head	Rota Kops 2007	4 subjects; 1.5 T MRI with T1w; atlas-based AC using an attenuation template obtained from 10 normal patients (T-AC)	Max. difference between PET_{T-AC} and PET_{TXAC} was 9% in occipital and frontal cortex. Future: Investigate gender-specific templates for T-AC.	[12]
Torso	Beyer, 2008	10 whole-body patients; 1.5 T MRI with T1w; CT-MR histogram matching	MR-AC challenged by differences in patient positioning, lack of MR surface coils and co-registration accuracy. Future: improve patient preparation on CT and MRI for accurate patient alignment	[1]
Head, Torso	Hofmann 2008, Hofmann 2008b	Head: 3 patients scanned on PET/CT and 1.5T MRI with T1w; Pseudo-CT prediction via atlas/machine learning method using 17 co-registered MR-CT pairs as training data. Torso: 5 whole-body patients; 1.5T MRI with T1w, (ongoing study)	Head: mean absolute difference PET_{MRAC} to PET_{CTAC} was 3.2%, maximum error of 10% in meningioma next to skull. Torso: Initial results show principal feasibility of predicting bone attenuation from MR. Future: acquire more suitable higher resolution MR (head) and expand on whole-body database (torso)	[8, 13]
Torso	Martinez 2009	2 whole-body patients; 1.5 T MRI with Dixon fat/water segmentation; 4 class segmentation of MR	Patient 1: 4.6% decrease in one ROI, Patient 2: 2.3% average difference in six ROIs. Future: account for variable attenuation in lung	[15]

mu = linear attenuation coefficient at 511 keV; T1w = T1-weighted MRI sequence; TX = transmission scan; T-AC = template-based AC; Future = extension of the study-related research as proposed by the authors.

bined devices were not yet considered seriously for clinical use. These early studies focused on PET applications in neurology [14, 41]. With the considerations of clinical PET/MR prototypes several groups have proposed algorithms for extra-cranial MR-AC as well [3, 15].

MR-AC can be evaluated by comparing the PET images that are obtained following standard TX-AC (transmission scan-based attenuation correction) or CT-AC and MR-AC. In the following we will refer to these images as PET_{TXAC}, PET_{CTAC} and PET_{MRAC}, respectively. Both TX-AC and CT-AC have their shortcomings. Depending on the scan time, TX scans exhibit relatively high noise levels that can be detrimental for AC purposes. On the other hand, CT images have noise levels that are several orders of magnitude lower than those obtained from images acquired with standard transmission sources (TX). However, the mapping of CT-based Hounsfield units to 511 keV attenuation values can be incorrect, particularly in the case of non-organic materials such as metal implants. Despite these problems, both TX-AC and CT-AC are commonly used and accepted. In accordance with the literature we will present PET_{TXAC} or PET_{CTAC} as the gold standard against which PET_{MRAC} is compared. The comparison can be done visually or quantitatively by means of relative differences of the reconstructed PET activity distributions. Differences can be assessed on a voxel-by-voxel basis, or, perhaps, more commonly for regions-of-interest (ROI).

ROIs can be defined automatically or manually by a clinical expert. For a study with n patients, where p ROIs are defined for each patient, it is impractical to quote all n × p differences. Therefore, it is preferred to report either the maximum differences or the mean absolute difference across all voxels. Some authors have quoted the mean differences, where the mean was taken from the positive or negative differences. This value indicates only the existence of an overall bias in the method, a value of zero for the differences would merely indicate that activity was overestimated as often as it was underestimated. Table 11.1 summarizes the results of the most significant studies on MR-AC.

11.2   MR-AC for brain applications

11.2.1   Segmentation approaches

MR-AC for brain applications was first addressed by Le Goff-Rougetet et al., who proposed a method to calculate PET AC factors from MR images in clinical examinations when both PET and MRI were required [14]. They argued that MR-AC helps simplify the clinical protocol and reduce the patient dose from standard PET transmission scanning. The methodology of their work, which they first applied to an FDG/water-filled cylindrical Lucite phantom, is based on a co-registration of the MR images to the PET transmission images using a surface matching technique. The co-registered MR images are then segmented into 3 classes (Table 11.1). Air was considered to be present only outside the patient. Appropriate linear attenuation coefficient values at 511 keV are then assigned to these tissue classes. Le Goff-Rougetet et al. evaluated their MR-AC method for phantoms and only 1 patient scan. Using ROIs within selected axial images of the phantom the authors found a maximum difference of 11% between PET_{MRAC} and PET_{TXAC}. The patient study revealed a maximum difference of 12%, primarily in the occipital cortex.

In a publication by El Fakhri and colleagues [12] MR-AC was also mentioned. However, the authors provided neither details of their implementation nor a performance evaluation. Personal communication with the authors suggests that they acquired 2 MR sequences for each subject and performed a cluster identification on the joint histogram prior to assigning the corresponding attenuation values.

An alternative method for MR-AC in brain PET was suggested by Zaidi et al. [41]. The authors had previously shown that the quality of PET neurology imaging was insufficient when standard PET attenuation correction methods were applied [40]. Therefore, the authors studied the use of MR-AC in brain PET (Table 11.1). They presented a workflow based on the availability of co-registered PET images, following standard (ellipse-fitted) attenuation correction, and MR images [41]. Using a fuzzy-logic-based segmentation method, the co-registered MR images were segmented into 5 tissues classes that were assigned attenuation coefficients at 511 keV. The entire process for MR-AC was reported to take 10 min on a Sun SPARC with minimal user intervention. Similar to Le Goff-Rougetet, the authors accounted for the head holder before using the segmented MR-based attenuation map for MR-AC.

Figure 11.2 shows brain images of an FDG-PET study from Zaidi et al [41] who compared segmented MR-AC with a standard, ellipse-fitted AC. The image quality of the PET following MR-AC appears somewhat improved, which can be attributed to the lower noise levels in the MR-based attenuation maps (Figure 11.2B). Analysis of the differences of the two methods of AC was performed across 10 patient sets. Despite a tendency of the method to lead to activity overestimation, overall correlation of ROI activity values on PET_{MRAC} and PET_{TXAC} was good (R^2=0.91), indicating the feasibility of segmented MR-AC as suggested by the authors (Table 11.1).

11.2.2   Atlas approaches

A viable alternative to multi-step segmentation procedures [41, 10] is to use atlas-registration (Figure 11.3). An atlas typically consists of a template MR image together with a corresponding attenuation label image. The template MR image can be obtained as an average of registered MR images from several patients. The label image could represent a segmentation into different tissue classes (e.g., air, bone, and soft tissue) or a co-registered attenuation map from a PET transmission scan or a CT scan with continuous attenuation values.

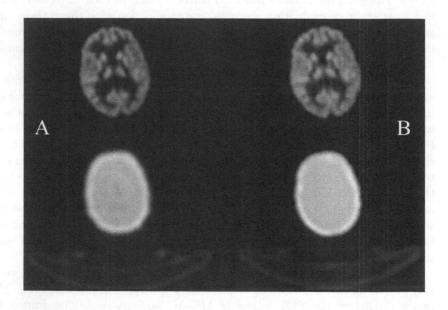

FIGURE 11.2: MR-AC for brain PET [41]. Axial slices through FDG-PET of the cerebellum of a patient following standard TX-AC (A) and MR-AC (B) with PET on top and the corresponding attenuation maps on the bottom. PET images appear visually similar with a slightly better signal-to-noise ratio following MR-AC.

Matched MR-CT Pair = Atlas

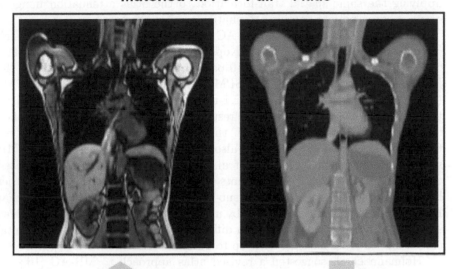

Determine registration
Atlas-MR to Patient-MR → *Apply registration to Atlas-CT*

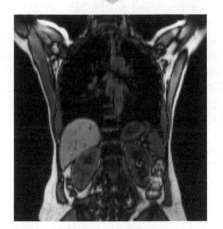
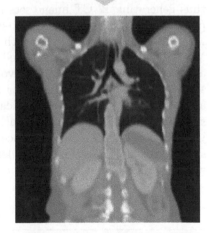

Patient MR   **MR-derived Pseudo-CT**

FIGURE 11.3: Principle of atlas-based MR-AC. The atlas consists of matching MR-CT image volumes that can be generated from patients. For a specific patient, the atlas MR image (top left) is co-registered to the MR image volume of the patient (bottom left). This transformation is then applied to the corresponding CT-atlas, thus generating an attenuation image (i.e., pseudo-CT) that matches the patient anatomy.

The template MR is warped to the patient-specific MR image volume. When applying the same spatial transformation to the atlas attenuation image a corresponding patient-specific attenuation map is generated.

Atlas-based approaches to MR-AC were presented by Rota Kops et al. [33] and Hofmann et al. [16]. Rota Kops and colleagues generated a template of PET transmission images from 10 patient datasets that is matched to the PET transmission template within SPM2 [13]. The MR template within SPM2 (which is already aligned with the PET transmission template) was normalized to the MR image of the patient. The resulting transformation was then applied to the template attenuation image, thus yielding an attenuation image for this patient. The same group has also employed MR-AC based on an MR segmentation method by Dogdas et al. [10]. Rota Kops et al. [33] validated their MR-AC algorithm with 4 patients (Figure 11.4). An analysis of the rather large ROIs defined on cortical and sub-cortical structures demonstrated that PET_{MRAC} differed from PET_{TXAC} by up to 10%. When using the MR-derived segmentation by Dogdas, maximum differences were observed in the occipital cortex and caudate nucleus, with up to 10% difference from PET_{TXAC}.

Hofmann et al. suggested a revised atlas approach to MR-AC [16]; see Figure 11.5. Here, the authors utilize a set of aligned MR-CT image volumes of 17 patients. Each of the available 17 MR image volumes from the MR-CT pairs was co-registered to the MR image volume from the PET/MR study. The co-registration vectors were applied to the corresponding CT image volumes thus generating 17 CT image sets that were aligned to the MR set from the patient. Subsequently, a pattern recognition approach was used to match the MR image of the patient with the appropriate CT information from that MR-CT data set that best matched the patient information. This voxel-based approach can merge partial sub-volumes from independent data sets into a single CT-volume that is used for MR-AC of the patients. This atlas-based algorithm was validated on 3 clinical data sets comparing MR-AC to the gold standard CT-AC [16]. Automated ROI analysis of PET_{MRAC} and PET_{CTAC} yielded a mean absolute difference of $3\% \pm 2.5\%$. Mean differences for the standard brain regions were smaller 10% and a maximum difference of 10% was observed for a meningioma located directly next to the skull.

11.3   Methods for torso imaging

Due to the current lack of prototype systems for a whole-body PET/MR system, studies of MR-AC algorithms for extra-cranial applications are scarce. Beyer et al. studied 10 patients who underwent routine torso scans (with arms up) on a combined PET/CT tomograph [3]. Within 1 day of the PET/CT exam, complementary MR scans were acquired. MR imaging was performed on a 1.5 T system with patients being positioned with their arms down.

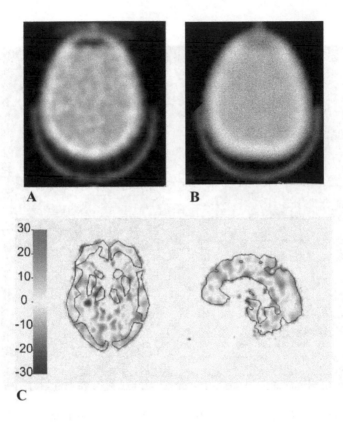

FIGURE 11.4: (See color insert.) MR-AC for brain PET: atlas-based attenuation correction. (A) Attenuation map measured through a PET transmission scan; (B) Attenuation map obtained through atlas registration and addition of the head holder; (C) Coronal and saggital view of voxel-by-voxel relative differences between PET attenuation corrected using attenuation maps A and B. (From [33].)

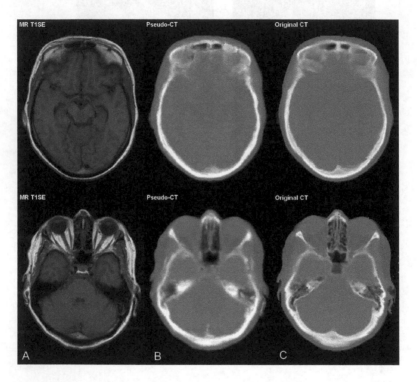

FIGURE 11.5: MR-AC for brain PET using the atlas-based method by Hofmann et al. [16]. Axial slices through patient data with mid-plane sections on top and lower brain sections on the bottom. (A) T1w spin-echo MR, (B) pseudo-CT as predicted from the atlas-based MR-AC and (C) original CT image. Note, the visual similarity between the pseudo-CT and the original CT.

Single-station, transverse T1-weighted VIBE MR images were used to generate pseudo-CT images. First, the MR images were co-registered to the CT images using non-linear, curvature-regularized co-registration in conjunction with mutual information. Second, the MR voxel value intensity distribution was matched to that of the co-registered CT image. MR-CT intensity transformation was performed in a 3-step process based on a histogram-matching algorithm. PET images were reconstructed on the PET/CT console following attenuation correction based on CT transmission images (PET_{CTAC}) and MR-based, pseudo-CT images (PET_{MRAC}); see Figure 11.6. The authors demonstrated that histogram matching is a feasible technique to transform MR to pseudo-CT attenuation values if the MR image quality is high. The study illustrated the need for accurate patient positioning between the MR and PET, but did not further quantify such effects.

Martinez-Möller et al. [23] proposed an approach to MR-AC where the attenuation map is segmented into background, lungs, fat, and soft tissue, which can be clearly delineated on MRI. The authors then evaluated the effect of "ignoring" bone tissue. Their study included 35 patients who had received ^{18}F-FDG PET/CT. On 52 lesions they used a CT-derived attenuation map that was segmented into the above four tissue classes, resulting in average SUV differences of $8\% \pm 3\%$ (mean $\pm$ SD) for $n = 21$ bone lesions, $4\% \pm 2\%$ for $n = 16$ neck lesions, and $2\% \pm 3\%$ for $n = 15$ lung lesions. The largest SUV difference was an underestimation of 13.1% for a lesion in the pelvic bone.

The authors then applied the Dixon segmentation method [9] as a proof of concept that such a 4-class attenuation map can be derived from an MR image (Figure 11.7). The Dixon segmentation was complemented by a component analysis to detect the lungs, and a morphological closing filter to avoid clas-

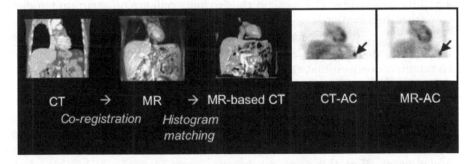

FIGURE 11.6: MR-AC for torso applications [3]. From top to bottom: CT images from PET/CT studies are co-registered to available MR. CT-MR histogram matching yields images with pseudo-CT attenuation values that are used for MR-AC. PET images following CT-AC and MR-AC show severe differences if the MR images inherit artifacts from suboptimal imaging protocols. In case of good MR image quality (see thorax) and accurate co-registration MR-AC based on histogram matching appears feasible.

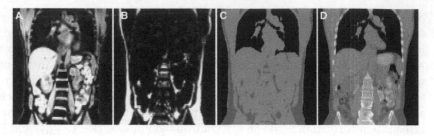

FIGURE 11.7: Segmentation of 2-point Dixon sequence[23]. MRI water (A) and fat (B) images are combined and segmented to generate a 4-class MRI-based attenuation map that does not include bone (C). (D) CT-based attenuation map of the same patient.

sifying bone voxels as air. Only two patients underwent both PET/CT and MRI examinations. Following MR-AC based on the Dixon method (ignoring bone) one lesion in the first patient showed an SUV underestimation of 4.6% and six lesions in the second patient had an SUV difference ranging from −6% to +1%.

Although predominantly used to date for brain imaging, atlas-based methods can also be applied to whole body imaging, allowing in principle to generate attenuation maps that include bone attenuation values. However, anatomic variability is high and it is unlikely that a general, spatial transformation captures all variables between a template and patient-specific anatomy. Hofmann et al. [16] presented a machine learning approach that combines the information from an atlas registration with local information that is drawn from small image patches. The method thus reduces reliance on accurate template to patient registration.

Even in cases where it is impossible to acquire an attenuation image through a transmission scan (or from an MR image), it is, in principle, possible to simultaneously estimate the transmission and emission image from emission data only [6, 29]. This was shown already in 1979 by Censor et al. [6]. Among others, Nuyts et al. [29] have further advanced this idea and incorporated it into a maximum-a-posteriori (MAP) algorithm. This allowed them to include additional prior knowledge about the attenuation coefficients (which usually fall into only a small number of classes) and about local smoothness. Despite a significant effort this approach has not been adopted widely. One of the reasons for this may be the artifacts that may arise from cross-talk between emission and attenuation image.

In very recent work, Salomon [35, 34] and colleagues have presented an approach that iteratively estimates the attenuation and emission images. The approach is based on a segmentation into anatomical regions (which could for example be derived from the MR image) and then uses PET emission data and consistency conditions to estimate attenuation coefficients for each segment.

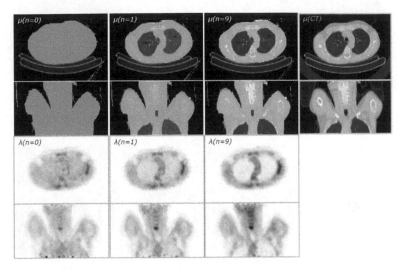

FIGURE 11.8: Simultaneous reconstruction of activity and attenuation distribution [35]: A study with patient data from PET/CT shows convergence of the estimated attenuation μ (upper rows) and activity λ (lower rows). Overall there is a good match between segmented (by simple region growing) estimated attenuation map and the CT-derived attenuation map.

Early results (Figure 11.8) show that in principle this allows to predict bone in the attenuation map. For clinical application no quantitative evaluation of this approach has yet been presented.

11.4 Discussion

Various approaches for predicting the attenuation maps from MR images on PET/MR examinations of patients have been reviewed. While segmentation-based approaches work well for brain applications, torso imaging with PET/MR may require more sophisticated methodologies, such as atlas-based image transformations from MR to pseudo-CT images. In general, MR-AC needs to address more than adequate transformation of MR pixel value information to appropriate PET attenuation values. In order for MR-AC to be viable in clinical routine it needs to account for additional challenges in torso and whole-body imaging. These challenges include the accurate representation of bone (typically not seen on MR), potential truncation effects from patients extending beyond the transverse field-of-view of the MR and the presence of MR surface coils typically not seen on MR.

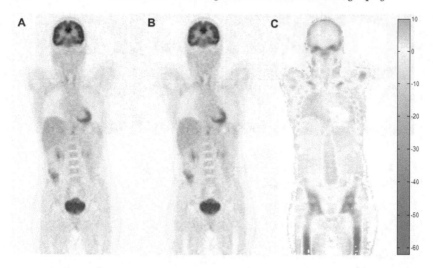

FIGURE 11.9: (See color insert.) Effect of ignoring cortical bone during AC. (A) PET image reconstructed using the original CT image. (B) PET image reconstructed using the same CT image with all bone structures set to the HU value of soft tissue thus simulating a best case scenario of MR-AC where bone attenuation is ignored. (C) Relative difference (%) between (A) and (B) illustrating the largest effect inside the skeleton. Note, voxels set to white in low uptake regions with SUV<0.2 in the original PET image.

11.4.1   The presence of bone

As bone structures are difficult to separate on MR images, a straightforward approach to MR-AC would be simply to ignore bone. This is not new and was shown in early studies on CT-AC to be of less impact than originally expected [1] despite the fact that the fraction of cortical bone varies in axial images, whereby it is higher in the head than in areas below the neck.

Figure 11.9 shows an example where CT-AC is performed with and without consideration of bone. This example illustrates the minimum bias expected from ignoring bone attenuation and considering this tissue class as soft tissue. In practice, an MR-AC algorithm that ignores bone tissue may also falsely attribute air as soft tissue and thus introduce much higher errors.

Figure 11.9 shows errors of up to 60% in the reconstructed PET image if bone structures in the attenuation map have been set to the attenuation value of soft tissue. This magnitude bias typically occurs in regions of relatively low activity, that are of less clinical interest.

The study by Martinez-Möller et al. [23] demonstrated no clinical difference in PET images following CT-AC and a (CT-derived) 4-tissue segmentation AC. Personal communication with nuclear medicine experts confirmed that quantitative errors of around 10%, or even 15% in lesion activity would typically not affect their diagnosis. However, larger studies, focussing on bone

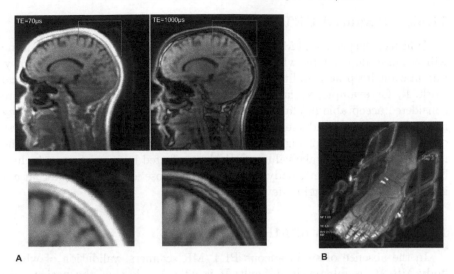

FIGURE 11.10: MR images acquired with a 3D ultrashort echo (UTE) sequences. (A) Sagittal brain section acquired with short echo: 0.07 ms (left) and late echo: 1 ms (right). (B) Angled view of a human foot. Note, bone yields a high signal and so do parts of the coil housing.

lesions and body parts with high bone contents, need to show whether ignoring bone is a clinically viable option.

Not accounting for bone in MR-based PET-AC is not an option for quantitative brain imaging: Receptor studies for example require an accuracy of less than 5% [7].

11.4.2 MR imaging with ultrashort echo time (UTE)

Instead of performing advanced image segmentation methods on standard MR images one may utilize dedicated MR sequences, such as ultrashort echo time (UTE) sequences [39, 32] that yield signal even from cortical bone (Figure 11.10). Typically, the use of just a single UTE image does not enable one to distinguish bone from non-bone tissues. However, when combined with a late echo image it is, in principle, possible to detect bone as the structure that yields a signal on the short echo image, but not on the late echo image (Figure 11.10A-B). By using multi-echo sequences [11] the 2 images can be acquired in one scan. While it seems promising for brain applications, UTE may not be acceptable as part of whole-body imaging protocols since acquisition time is on the order of several minutes per bed position.

11.4.3   Required PET accuracy

Which of the presented MR-AC methods will be accepted in clinical routine will not only depends on what MR-AC method yields the highest accuracy, but also on its practicability and on what accuracy is sufficient in clinical work. If, for example, variations of up to 10% in PET activity values are considered acceptable in clinical routine, then methods with higher accuracy might be dismissed and methods with other advantages such as robustness or computational speed might be preferred. PET uptake values, such as the measured standard uptake values (SUV), are affected by many factors including length of uptake time, body composition, glucose load and others that are independent of the imaging device [18].

11.4.4   Validation of MR-AC methods

In the absence of simultaneous PET/MR scanners, validation of whole-body MR-AC is inherently difficult: It is unavoidable that the patient will move between MR and PET(CT) examinations. Therefore, even if the attenuation map could be predicted with a high accuracy from the MR image, the patient movement between the MR and a TX, or CT scan may still cause the MR-predict attenuation map to be different from the reference scan attenuation values, thereby also causing a difference between PET_{MRAC} and PET_{CTAC}. Several authors [3] have suggested compensating for the patient motion between scans by performing non-rigid MR-CT co-registration. It should, however, be noted that not all motion-related misalignments can be corrected. For example, pockets of gas in the abdominal region may vary significantly between scans, or not even appear on one of the 2 complementary exams. The validation of non-linear co-registration algorithms remains an open issue [30] that requires addressing.

11.4.5   Truncated field-of-view

In clinical PET/MR imaging, patients may well extend beyond the transverse FOV of the MR. Thus, the arms or even the trunk of the patient are not fully covered by the MR image. Nonetheless, the contribution of the truncated anatomy to overall attenuation needs to be accounted for. The very same problem was described for PET/CT applications where truncated attenuation maps were shown to yield significant image distortion and bias near the area of truncation [24, 38, 2]. Recently, Delso and colleagues discussed the effect of MR truncation on MR-AC [8]. Their study showed that when the arms were outside the FOV the PET activity after AC was biased by up to 14% in that area. Using simple image processing techniques they could recover the arms in the truncated image and thus reduce the quantification bias to 2%. An alternative solution would be to use the uncorrected PET image to estimate the patient cross-section in those areas outside the measured FOV where no MR information is available. The feasibility of such an approach still needs to be validated. In imaging scenarios with highly specific tracers, the arms may be difficult to segment automatically in the uncorrected PET images. Yet another approach could predict the body cross-section, including parts which were outside the field of view of the MR image, from the atlas registration. In theory, these approaches could even be combined such that the registration is performed based on the MR image where the MR image is available, and elsewhere based on the uncorrected PET image.

11.4.6 MR coils and positioning aids

The fact that the MR coils are located inside the FOV of the PET system is a challenge for MR-AC. For brain scans, the head coil is rigid and its attenuation values can be estimated from a baseline CT. Subsequently for any PET/MR study knowledge of the relative position of the head coil inside the PET/MR system would be required. Similarly other rigid objects inside the PET system could be detected and included in the MR-based attenuation map [22].

For extra-cranial examinations the situation is far more difficult. Surface coils are required to avoid suboptimal signal generation. Surface coils contain elastic components and hence cannot be located easily with respect to the gradient coil or the patient. MR sequences with UTE could possibly help detect landmarks of surface coils and thus help account for their attenuation. However, in any case for PET/MR systems the RF coils need to be designed such that low photon attenuation and scatter are ensured.

Positioning aids are commonly used to reduce patient movement over the duration of the scan. As they are typically made of foam or other soft materials that can be adapted to the patient outline, their shape and corresponding attenuation map changes from one patient scan to the next. This position seems difficult to predict even if MR-visible markers were placed on them. Mantlik et al. [21] recently performed a quantitative analysis of the effect of ignoring the attenuation of positioning aids. For 5 head/neck patients who were fixed in a vacuum mattress they obtained a mean underestimation of activity of 9.1% if the mattress was not included in the attenuation map.

11.4.7 User intervention

Ideally, for the application in clinical PET/MR scenarios, MR-AC should be fully automatic in order to limit user interaction and, subsequently, examination and processing times. Despite claims of some groups that their method for MR-AC is "robust," problems remain that require "some manual intervention of the operator" [41]. Thus, automation of MR-AC remains a challenge, particularly in cases of large deviation from normal anatomy.

11.4.8   Potential benefits of MR-AC

In PET/CT the PET image is acquired over several minutes, while the CT scan is a matter of seconds and frequently acquired during a single breath hold. As a result, patient or organ motion typically causes local misalignment between the PET and CT images and may lead to serious artifacts for AC, for example near the diaphragm [4]. Some authors have recommended 4D PET/CT acquisition and AC [27, 31], however, this involves a substantially higher patient radiation dose. As MR scans generally take much longer than CT scans, patients conceivably spend an even longer time in the PET/MR scanner compared to PET/CT and consequently patient motion is likely to cause even more severe artifacts. Here, the use of periodic MR navigator signals in conjunction with a 4D model of the human torso may help to correct for motion-induced image degeneration in PET/MR data following 4D-MR-AC, which would be a major advantage over CT-AC.

11.4.9   Additional potential benefits of simultaneous PET/MR acquisition

As early as 1991, Leahy et al. [19] suggested that PET reconstruction could be improved by using anatomical MR images from the same patient as prior information. This remains a field of active research and the potential of the method can be seen in Figure 11.11. While it is commonplace today that almost all neurology patients who receive a PET scan also receive a MR scan, MR-guided PET reconstruction has not yet made the transition from research into clinical routine. Aside from logistical problems of automatically retrieving the matching MR image from the PACS, one of the reasons for this might be that mis-registrations, which are unavoidable in retrospective PET-MR co-registration, quickly deteriorate the image quality [20]. In combined PET/MR tomographs, the co-registration accuracy is improved and may help promote the concept of MR-guided PET image reconstruction.

Even if the PET image is reconstructed independently of the MR image, it is still possible to use the MR image of the patient as an aid for improved quantification. In particular MR-guided partial volume correction (PVC) was suggested as early as 1990 [26, 25]. Again, PET and MR images from combined PET/MR examinations may facilitate improvements in MR-based PET quantification through the use of MR-based PVC.

11.5   Conclusion

With the onset of a research interest in combined PET/MR imaging several studies have appeared on the use of MR for AC of the PET data. MR-AC

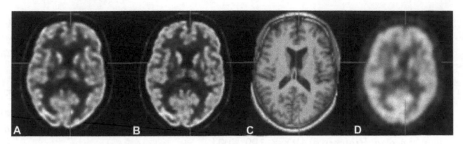

FIGURE 11.11: (A) Maximum Likelihood (ML) PET reconstruction that only used the PET raw data, (B) MAP-Reconstruction that used a joint entropy prior based on an MRI image of the same patient, (C) MRI, (D) Standard Filtered Back Projection Reconstruction. Image with permission from Nuyts et al. [28]

is not as straightforward as CT-AC that allows the estimation of 511 keV attenuation maps from CT transmission images. In the absence of CT-like transmission sources in PET/MR, alternative solutions to MR-AC include the use of complex segmentation tools that were shown to work for brain applications. In extra-cranial PET/MR the percentage of bone voxels is significantly smaller and an accuracy sufficient for many applications might be achieved by simply "ignoring" bone, provided that the algorithm does not falsely predict air at bone locations. If higher accuracy is required, other approaches that include atlas-based methods and simultaneous reconstruction of attenuation and activity distributions allow predicition of bone attenuation and appear more promising. While MR-AC is work-in-progress, further advantages of MR-AC over CT-AC become apparent, which include the additional use of MR for retrospective motion correction or partial volume correction of the PET.

References

[1] T. Beyer, P.E. Kinahan, D.W. Townsend, and D. Sashin. The use of x-ray CT for attenuation correction of PET data. In *IEEE Conference Record of the Nuclear Science Symposium and Medical Imaging Conference*, volume 4, page 1573, 1994.

[2] T. Beyer, M. Weigert, C. Palm, H. Quick, S. Müller, U. Pietrzyk, F. Vogt, M.J. Martinez, and A. Bockisch. Towards MR-based attenuation correction for whole-body PET/MR imaging. In *Society of Nuclear Medicine Annual Meeting Abstracts*, volume 47, Supplement 1, page 384, 2006.

[3] T. Beyer, M. Weigert, H.H. Quick, U. Pietrzyk, F. Vogt, C. Palm, G. Antoch, S.P. Müller, and A. Bockisch. MR-based attenuation correction for torso-PET/MR imaging: pitfalls in mapping MR to CT data. *European Journal of Nuclear Medicine and Molecular Imaging*, 2008.

[4] A. Bockisch, T. Beyer, G. Antoch, L.S. Freudenberg, H. Kühl, J.F. Debatin, and S.P. Müller. Positron emission tomography/computed tomography—imaging protocols, artifacts, and pitfalls. *Molecular Imaging and Biology*, 6(4):188–199, 2004.

[5] C. Catana, D. Procissi, Y. Wu, M. S. Judenhofer, J. Qi, Bernd J. Pichler, R. E. Jacobs, and S. R. Cherry. Simultaneous in vivo positron emission tomography and magnetic resonance imaging. *Proceedings of the National Academy of Science* USA, 11:3705–3710, 2008.

[6] Y. Censor, D.E. Gustafson, A. Lent, and H. Tuy. A new approach to the emission computerized tomography problem: simultaneous calculation of attenuation and activity coefficients. *IEEE Transactions on Nuclear Science*, 26(8 Part 2):2775–2779, 1979.

[7] R. De La Fuente-Fernandez, J.Q. Lu, V. Sossi, S. Jivan, M. Schulzer, J.E. Holden, C.S. Lee, T.J. Ruth, D.B. Calne, and A.J. Stoessl. Biochemical variations in the synaptic level of dopamine precede motor fluctuations in Parkinson's disease: PET evidence of increased dopamine turnover. *Annals of Neurology*, 49(3):298–303, 2001.

[8] G. Delso, R. Bundschuh, A. Martinez-Möller, S. Nekolla, S. Ziegler, and M. Schwaiger. Impact of limited MR field-of-view in simultaneous PET/MR acquisition. In *Society of Nuclear Medicine Annual Meeting Abstracts*, volume 49, Supplement 1, page 162P. Soc Nuclear Med, 2008.

[9] W.T. Dixon. Simple proton spectroscopic imaging. *Radiology*, 153(1):189, 1984.

[10] B. Dogdas, D. W. Shattuck, and R. M. Leahy. Segmentation of skull and scalp in 3-D human MRI using mathematical morphology. *Human Brain Mapping*, 25:273–285, 2005.

[11] J. Du, M. Bydder, and A. Takahashi. Two-dimensional ultrashort echo time imaging using a spiral trajectory. *Magnetic Resonance Imaging*, 2007.

[12] G. El Fakhri, M.F. Kijewski, K.A. Johnson, G. Syrkin, and R.J. Killiany. MRI-guided SPECT perfusion measures and volumetric MRI in prodromal Alzheimer disease. *Archives of Neurology*, 60:1066–1072, 2003.

[13] K.J. Friston, J. Ashburner, C.D. Frith, J.B. Poline, J.D. Heather, and R.S.J. Frackowiak. Spatial registration and normalization of images. *Human Brain Mapping*, 2(1):1–25, 1995.

[14] R. Le Goff-Rougetet, V. Frouin, J.-F. Mangin, and Bernard Bendriem. Segmented MR images for brain attenuation correction in PET. In *Proceedings of SPIE*, pages 725–736, 1994.

[15] M. Hofmann, F. Steinke, P. Aschoff, M. Lichy, J.M. Brady, B. Schoelkopf, and B.J. Pichler. MR-based PET attenuation correction initial results for whole body. In Abstract at *IEEE Nuclear Science Symposium and Medical Imaging Conference*, 2008.

[16] M. Hofmann, F. Steinke, V. Scheel, G. Charpiat, J. Farquhar, P. Aschoff, M. Brady, B. Schölkopf, and B. J. Pichler. MR-based attenuation correction for PET/MR: a novel approach combining atlas registration and recognition of local patterns. *Journal of Nuclear Medicine*, 49(11):1875–1883, 2008.

[17] M.S. Judenhofer, H.F. Wehrl, D.F. Newport, C. Catana, S.B. Siegel, M. Becker, A. Thielscher, M. Kneilling, M.P. Lichy, M. Eichner, et al. Simultaneous PET-MRI: a new approach for functional and morphological imaging. *Nature Medicine*, 14(4):459–465, 2008.

[18] J.W. Keyes Jr. SUV: standard uptake or silly useless value? *Journal of Nuclear Medicine*, 36(10):1836–1839, 1995.

[19] R.M. Leahy and X. Yan. Incorporation of anatomical MR data for improved functional imaging with PET. *Proceedings of 12th International Conference on Information Processing in Medical Imaging*, pages 105–120, 1991.

[20] B. Lipinski, H. Herzog, E. Rota Kops, W. Oberschelp, and H.W. Muller-Gartner. Expectation maximization reconstruction of positron emission tomography images using anatomical magnetic resonance information. *IEEE Transactions on Medical Imaging*, 16(2):129–136, Apr 1997.

[21] F. Mantlik, M. Hofmann, J. Kupferschläger, M. K. Werner, B. J. Pichler, and T. Beyer. The effect of positioning aids on PET quantification following MR-based attenuation correction (AC) in PET/MR imaging. *Journal of Nuclear Medicine*, 51 (Supplement 2):1418, 2010.

[22] D. Martin, G. Platsch, M. Requardt, S. Schmidt, K. Schmiedehausen, and M. Szimtenings. Method for determining attenuation values for PET data of a patient. US Patent Application 20090105583, 2009.

[23] A. Martinez-Möller, M. Souvatzoglou, G. Delso, R. A. Bundschuh, C. Chefd'hotel, S. I. Ziegler, N. Navab, M. Schwaiger, and S. G. Nekolla. Tissue classification as a potential approach for attenuation correction in whole-body PET/MRI: evaluation with PET/CT data. *Journal of Nuclear Medicine*, 50(4):520–526, 2009.

[24] O. Mawlawi, J.J. Erasmus, T. Pan, D.D. Cody, R. Campbell, A.H. Lonn, S. Kohlmyer, H.A. Macapinlac, and D.A. Podoloff. Truncation artifact on PET/CT: impact on measurements of activity concentration and assessment of a correction algorithm. *American Journal of Roentgenology*, 186(5):1458–1467, 2006.

[25] C.C. Meltzer, P.E. Kinahan, P.J. Greer, T.E. Nichols, C. Comtat, M.N. Cantwell, M.P. Lin, and J.C. Price. Comparative evaluation of MR-based partial-volume correction schemes for PET. *Journal of Nuclear Medicine*, 40(12):2053–2065, 1999.

[26] C.C. Meltzer, J.P. Leal, H.S. Mayberg, H.N. Wagner Jr, and J.J. Frost. Correction of PET data for partial volume effects in human cerebral cortex by MR imaging. *Journal of Computer Assisted Tomography*, 14(4):561, 1990.

[27] S.A. Nehmeh, Y.E. Erdi, T. Pan, A. Pevsner, K.E. Rosenzweig, E. Yorke, GS Mageras, H. Schoder, P. Vernon, O. Squire, et al. Four-dimensional (4D) PET/CT imaging of the thorax. *Medical Physics*, 31:3179, 2004.

[28] J. Nuyts. The use of mutual information and joint entropy for anatomical priors in emission tomography. *IEEE Nuclear Science Symposium and Medical Imaging Conference Record*, 6:4149–4154, 26 2007-Nov. 3 2007.

[29] J. Nuyts, P. Dupont, S. Stroobants, R. Benninck, L. Mortelmans, and P. Suetens. Simultaneous maximum a posteriori reconstruction of attenuation and activity distributions from emission sinograms. *IEEE Transactions on Medical Imaging*, 18(5):393–403, 1999.

[30] U. Pietrzyk. Does PET/CT render software registration obsolete? *Nuklearmedizin*, 44:13–17, 2005.

[31] F. Pönisch, C. Richter, U. Just, and W. Enghardt. Attenuation correction of four dimensional (4D) PET using phase-correlated 4D-computed tomography. *Physics in Medicine and Biology*, 53(13):N259–N268, 2008.

[32] M.D. Robson, P.D. Gatehouse, M. Bydder, and G.M. Bydder. Magnetic resonance: an introduction to ultrashort TE (UTE) imaging. *Journal of Computer Assisted Tomography*, 27(6):825–846, 2003.

[33] E. Rota Kops and H. Herzog. Alternative methods for attenuation correction for PET images in MR-PET scanners. *IEEE Nuclear Science Symposium and Medical Imaging Conference Record*, 6, 2007.

[34] A. Salomon and A. Goedicke. Apparatus and method for generation of attenuation map, May 14 2009. WO Patent WO/2009/060,351.

[35] A. Salomon, V. Schulz, R. Brinks, B. Schweizer, A. Goedicke, and T. Aach. Iterative generation of attenuation maps in TOF-PET/MR

using consistency conditions. In *Society of Nuclear Medicine Annual Meeting Abstracts*, volume 50, Supplement 2, page 2013. Soc. Nuclear Med., 2009.

[36] H.-P. Schlemmer, B. Pichler, K. Wienhard, M. Schmand, C. Nahmias, D. Townsend, W.-D. Heiss, and C. Claussen. Simultaneous MR/PET for brain imaging: first patient scans. *Journal of Nuclear Medicine Meeting Abstracts*, 48:45, 2007.

[37] Y. Shao, S. R. Cherry, K. Farahani, K. Meadors, S. Siegel, R. W. Silverman, and P. K. Marsden. Simultaneous PET and MR imaging. *Physics in Medicine and Biology*, 10, 1997.

[38] E. Tsukamoto and S. Ochi. PET/CT today: system and its impact on cancer diagnosis. *Annals of Nuclear Medicine*, 20(4):255–67, 2006.

[39] A. Waldman, J.H. Rees, C.S. Brock, M.D. Robson, P.D. Gatehouse, and G.M. Bydder. MRI of the brain with ultra-short echo-time pulse sequences. *Neuroradiology*, 45(12):887–892, 2003.

[40] H. Zaidi and B. Hasegawa. Determination of the Attenuation Map in Emission Tomography. *Journal of Nuclear Medicine*, 44:291–315, 2002.

[41] H. Zaidi, M.-L. Montandon, and D.O. Slosman. Magnetic resonance imaging-guided attenuation and scatter corrections in three-dimensional brain positron emission tomography. *Medical Physics*, 30:937–948, 2003.

Chapter 12

Optical Imaging

Angelique Ale and Vasilis Ntziachristos

Institute for Biological and Medical Imaging, Technical University of München and Helmholtz Center München, Neuherberg, Germany

12.1	Introduction		241
12.2	Fluorescence molecular tomography (FMT)		244
	12.2.1	Light propagation model	244
		12.2.1.1 Photon interaction with biological tissue	244
		12.2.1.2 The diffusion approximation	246
		12.2.1.3 Model for a fluorescence heterogeneity	248
	12.2.2	Reconstruction of the fluorochrome distribution	249
12.3	FMT and hybrid FMT systems		251
	12.3.1	Instrumentation	251
		12.3.1.1 Illumination	251
		12.3.1.2 Detection	252
		12.3.1.3 360° projections	252
	12.3.2	Multimodal optical imaging	253
		12.3.2.1 Optical tomography and MRI	253
		12.3.2.2 FMT-XCT	254
References			257

12.1 Introduction

Optical imaging refers to imaging techniques that use light as a tool of observation. A large variety of light sources, from visible to near-infrared light, can be used in many different source-detector configurations. This has enabled the development of a wide range of imaging methods and approaches from the macroscopic observation of a patient or an animal to microscopic studies, with numerous applications in clinical practice and biomedical research.

New optical imaging techniques are continuously emerging at the macroscopic [4, 14, 32, 41, 28] and microscopic levels [2, 19, 20, 21, 47, 50, 51, 60, 61]. Macroscopic developments include mainly the noninvasive imaging of functional and molecular contrast in tissues *in vivo* [34, 40, 58]. These approaches take advantage of the wealth of contrast mechanisms and physical properties

241

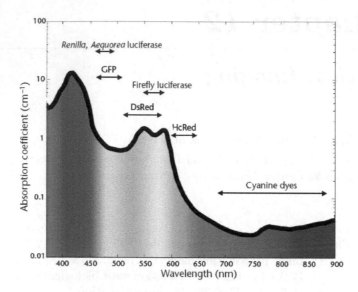

FIGURE 12.1: Absorption spectra of tissue and spectral emission range of common fluorescent probes and proteins. (From [58].)

of light that can be used to form images. Optical imaging technologies are well suited for experimentation, as most of the components required can be assembled on the laboratory bench and are modular in design. In addition, high quality images can be obtained at moderate cost, and the systems can be made portable or compact.

Insights on the functional and molecular levels can be obtained by visualization of endogenous tissue contrast, such as absorption and scattering properties. The observation of molecular events and processes can be facilitated by exogenous contrast using fluorescence or bioluminescence approaches. The recent increase in availability of fluorescent proteins, dyes and probes [27, 53, 10, 59, 33] has sparked interest in the combination of optical imaging with fluorescence. Fluorescence imaging can be divided into two main strategies, (a) direct imaging, in which an engineered fluorescent probe that localizes a specific target is introduced in the imaging subject, and (b) indirect imaging, where transgenic methods are used for intrinsic expression of fluorescence [31].

Tissue of several centimeters thickness can be imaged when near-infrared light is used, due to the low attenuation of light by tissue in this range; see Figure 12.1. Several different techniques have been developed for the macroscopic imaging of fluorescence *in vivo*. In epi-illumination (reflectance) imaging, the tissue surface is illuminated with an expanded light beam and images are collected from the same side of the tissue; see Figure 12.2. When the light propagates through the tissue, it excites superficial and subsurface

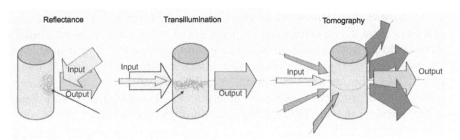

FIGURE 12.2: In fluorescence reflectance imaging, excitation light is expanded on the object surface and fluorescence light is collected from the same side of the object. In transillumination mode, illumination and detection are performed on opposite sides. In fluorescence tomography, the object is illuminated from different angles in transillumination mode. (From [34].)

fluorochromes. The light emitted by the fluorochromes is then captured by a camera using appropriate filters. An alternative method to epi-illuminant imaging is transillumination imaging, based on the same principle, but with source and detectors placed on opposite sides of the tissue. An advantage of transillumination is the larger feasible penetration depth. Due to the nonlinear dependencies of light propagation through tissue, significant uncertainty on the exact depth of the recorded signal exists with both methods. The depth of the signal can be more accurately resolved when tomographic imaging is used. In this case transillumination images from multiple source-detector configurations are recorded and combined to a three-dimensional reconstruction of the internal fluorochrome distribution. This technique is termed fluorescence molecular tomography (FMT) and is the main subject of this chapter.

The principle of operation in FMT resembles that of X-ray computed tomography (CT) in that tissue is illuminated from different angles and at different positions and a mathematical formulation is used to describe photon propagation in tissue. However, a major difference between optical tomography and tomographic methods based on high-energy rays is that photons in the optical range are highly scattered by tissue organelles and membranes. Photons do not propagate in straight lines when traveling through tissue, but become diffuse within a few millimeters of propagation. The diffusive nature of the light propagation through tissue limits the quantification ability and maximum resolution that can be achieved. Therefore, FMT is mainly concerned with the localization and quantification of bulk signals from specific fluorescent entities indicating cellular and molecular activity.

One of the most recent technological evolutions in the field of FMT has been the development of multimodality systems, in which FMT is combined with X-ray CT or MRI. A straightforward benefit of hybrid methods is that they allow the seamless co-registration of the images obtained, since all the modalities employed visualize the object of interest under identical placement.

Using hybrid three-dimensional visualization, different contrast mechanisms can be accurately superimposed, improving the information content available. An additional important benefit of the hybrid approach is that it enables the improvement of the performance and image quality of the optical method by utilizing the anatomical information from X-ray CT to form a map of tissue optical properties, or a map of tissue structural information, for input in the physical model. This strategy can improve the accuracy of the optical reconstruction problem [22, 1, 25].

This chapter summarizes several concepts important in fluorescence tomography. Section 12.2 outlines a common theoretical description of the physical model of light propagation through tissue, which is the basis for the reconstruction of the fluorochrome distribution. Section 12.3 gives an overview of several existing fluorescence tomography systems and expected progress of hybrid systems.

12.2   Fluorescence molecular tomography (FMT)

Optical tomography is generally based on a physical model of photon propagation and therefore not only yields three-dimensional imaging but also quantification of optical contrast. The different propagation regimes associated with optical imaging range from the ballistic regime (at low scattering conditions), where light propagation is described by the laws of geometrical optics, to the diffusive regime where the directivity is lost due to multiple scattering events [33]. Photon propagation in scattering media can be accurately described by the radiative transfer model [53, 49, 44, 5, 8, 7]. However, instead of the radiative transfer model, the diffusion approximation is more often used for biomedical imaging, due to the smaller degree of complexity and less intensive computational efforts associated with it. In the following, the diffusion model [33], as well as an approach for the reconstruction of a fluorescent source inside a diffuse medium [31], is outlined.

12.2.1   Light propagation model

12.2.1.1   Photon interaction with biological tissue

Two interactions of light with tissue play a major role in optical imaging: absorption and scattering of photons [13, 18, 26, 55, 57]. Absorption occurs when a photon incident on a molecule has an energy corresponding to the energy difference between two molecular orbital levels. The radiation field transfers its energy to the molecule and an electron is excited from the ground level to a higher-energy state; see Figure 12.3. Several secondary processes can follow absorption, such as the emission of new photons or heat production.

Three processes that do not involve the emission of new photons can occur, collectively called non-radiative decay: vibrational relaxation, which is energy tranfer through collisions with the surrounding material, intersystem crossing, which is a transition between two different spin-states, and internal conversion, a transition between two energy states with the same spin angular momentum. Emission of photons can be summarized in two processes: fluorescence, which is the emission from singlet-excited states and phosphorescence, the emission from triplet-excited states; see Figure 12.3. Scattering is the redirection of a photon after interaction with matter. In elastic or Rayleigh scattering, there is no energy exchange between photon and molecule and the scattered photon has the same wavelength as the incident photon. In inelastic or Raman scattering, energy transfer occurs, and the scattered photon has a different energy than the incident photon. Rayleigh scattering is dominant in tissue.

The above-described processes are captured in the light propagation model through the absorption coefficient μ_a [cm^{-1}] and scattering coefficient μ_s [cm^{-1}], defined by:

$$\mu_a = C_a \sigma_a, \mu_s = C_s \sigma_s, \tag{12.1}$$

where $C_{a,s}$ [cm^{-3}] are the concentrations of absorbers and scatterers in the medium, and $\sigma_{a,s}$ [cm^2] are the absorption and scattering cross section, respectively. Their reciprocal values represent the average attenuation length:

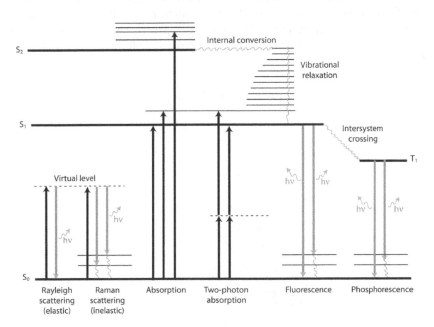

FIGURE 12.3: Jablonski diagram illustrating possible interactions of light with matter. S0, S1 and S2 are the ground, first and singlet electronic states, T1 is the first triplet state. (From [33].)

$1/\mu_a$ is the average distance covered by photons before being absorbed, and $1/\mu_s$ is the average distance between scattering events [7, 11, 12, 48, 52, 54].

In the following sections, we describe some key equations used in the to-mographic problem following the diffusion approach [33, 3]. Their purpose is to provide the reader a basic outline of formulating a fluorescence optical tomography problem.

12.2.1.2   The diffusion approximation

The diffusion model can be derived from the general principle of conservation of energy and Fick's law. The principle of conservation of energy for a diffusive medium applied over a finite volume can be stated as follows: the change of the energy density in the volume over time equals: − power density leaving the volume − power density absorbed in the volume + power density produced by sources present.

The energy density $w(\vec{r}, t)$ [J cm^{-3}] is the energy per unit volume of the radiation field, $w(\vec{r}, t) = dE/dV$, where V is the volume. The energy density is related to the fluence rate $\phi(\vec{r}, t)$ by:

$$w(\vec{r}, t) = \phi(\vec{r}, t)/c, \tag{12.2}$$

where c is the speed of light in the medium. The fluence rate $\phi(\vec{r}, t)$ [W cm^{-2}], also called spectral irradiance, is defined as the power incident on a small sphere at a given point $\vec{r}$ in space, divided by the cross-sectional area of that sphere. It can be written as the integral over all directions of the radiance $L(\vec{r}, \hat{s}, t)$:

$$\phi(\vec{r}, t) = \int \int_{4\pi} L(\vec{r}, \hat{s}, t) d\hat{s}. \tag{12.3}$$

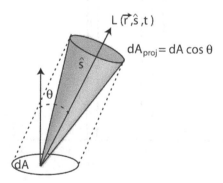

FIGURE 12.4: The radiance L is defined as the flux per unit projected area per unit solid angle leaving a source or a reference surface: $L = dP/d\hat{s}dA_{proj}$. (From [33].)

The radiance $L(\vec{r}, \hat{s}, t)$ [W cm^{-2} st^{-1}], see Figure 12.4, is defined as the flux per unit projected area per unit solid angle leaving a source or a reference surface: $L = dP/d\hat{s}dA_{proj}$, with P [W] the power or rate at which energy is transferred from one region to another by the radiation field, $dA_{proj} = dA\cos\theta$ is the projected area, and θ is the angle between the outward surface normal of the area element dA and the direction of observation $\hat{s}$. The flux density vector $\vec{j}(\vec{r}, t)$ equals the first moment of the radiance:

$$j(\vec{r}, t) = \int\int_{4\pi} L(\vec{r}, \hat{s}, t)\hat{s}d\hat{s} \qquad (12.4)$$

The irradiance I [W cm^{-2}] is defined as the flux per unit area received by a real or imaginary surface, $I = dP/dA$. The flux density vector is also related to the irradiance by:

$$I(\vec{r}, t) = j(\vec{r}, t)\cdot\hat{n}. \qquad (12.5)$$

The principle of conservation of energy can now be formulated in terms of the above-mentioned quantities as:

$$\underbrace{\int_V \frac{1}{c}\frac{\partial\phi(\vec{r}, t)}{\partial t}dV}_{change\ in\ V} = \underbrace{\int_{\partial V} j(\vec{r}, t)\cdot\hat{n}dS}_{flow} + \underbrace{\int_V (-\mu_a\phi(\vec{r}, t) + S(\vec{r}, t)dV}_{production} \qquad (12.6)$$

where c [cm s^{-1}] is the speed of light in the medium, $\phi(\vec{r}, t)$ [W cm^{-2}] is the fluence rate, $j(\vec{r}, t)$ [W cm^{-2}] is the flux density vector, $\hat{n}$ is the normal on the surface, μ_a [cm^{-1}] is the absorption coefficient and $S(\vec{r}, t)$ [W cm^{-3}] is the power per unit volume produced by the sources. After applying Gauss' theorem to the surface integral, Equation 12.6 reduces for an arbitrary volume to:

$$\frac{1}{c}\frac{\partial\phi(\vec{r}, t)}{\partial t} = -\nabla j(\vec{r}, t) - \mu_a\phi(\vec{r}, t) + S(\vec{r}, t). \qquad (12.7)$$

To solve Equation 12.7, another relationship between the fluence rate and the flux density vector is necessary. For highly scattering media, the photons follow random path trajectories. After a few scattering events the photons have a diffusive behavior. An accurate description of the diffusion process is provided by Fick's law [38]:

$$j(\vec{r}, t) \approx -D\nabla\phi(\vec{r}, t), \qquad (12.8)$$

with D the diffusion coefficient. The diffusion coefficient D depends on the absorption and scattering coefficients through [7]:

$$D = \frac{1}{3[\mu'_s + \alpha\mu_a]}, \qquad (12.9)$$

where μ'_s is the reduced scattering coefficient defined by $\mu'_s = \mu_s(1 - g)$, with g the anisotropy factor. Using this expression, the diffusion of photons can be approximated as an isotropic scattering process, while the actual anisotropy is captured by g. A medium scattering mostly in the forward direction, for which $g \approx 1$, is assumed equivalent to a medium having a smaller reduced scattering coefficient μ'_s. In the near-infrared range, $\mu'_s \gg \mu_a$, and it is customary to define the diffusion coefficient by:

$$D = \frac{1}{3[\mu'_s]}. \tag{12.10}$$

Substituting Fick's law in Equation 12.7 leads to the following diffusion equation:

$$\frac{1}{c}\frac{\partial \phi(\vec{r}, t)}{\partial t} = -\nabla D \nabla \phi(\vec{r}, t) - \mu_a \phi(\vec{r}, t) + S(\vec{r}, t). \tag{12.11}$$

For the time-independent case, the diffusion equation simplifies to

$$-\nabla D(\vec{r})\nabla \phi \vec{r}, t) + \mu_a \phi(\vec{r}, t) = S(\vec{r}, t). \tag{12.12}$$

The general dependence of the photon field on optical properties and source detector distance has the following form:

$$\phi(\vec{r}, t) \sim \frac{e^{ikr}}{r}. \tag{12.13}$$

Where r is the source-detector distance assuming a point source and a point detector, and $k = (\frac{-c\mu_a + i\omega}{cD})^{1/2}$ is the propagation wavenumber of the photon wave that depends on the absorption coefficient μ_a, the diffusion coefficient D, the speed of light c in tissue and the modulation frequency ω of the photon beam that illuminates the tissue. For light of constant intensity, $\omega = 0$. Equation 12.13 describes a generic dependence that does not account for the effects of heterogeneities or of boundaries, but illuminates the complex nature of photon attenuation in tissues.

12.2.1.3   Model for a fluorescence heterogeneity

Next to the general description of light traveling through scattering media, we need an expression for the light propagation when a fluorochrome is present in the medium [31]. Fluorescence heterogeneities in the medium can be treated similar to absorption or scattering heterogeneities in the medium. Here we will derive the corresponding expressions using the Born approximation [24]. We consider the total field $\phi(\vec{r})$ to be a sum of $\phi_0(\vec{r})$ and $\phi_{sc}(\vec{r})$, where $\phi_0(\vec{r})$ is the incident field, defined as the field without heterogeneities, and $\phi_{sc}(\vec{r})$ is the scattered field, defined as the field that can be attributed to the heterogeneities. We assume a distribution $O(\vec{r})$ of an optical property perturbation; this could be the distribution of the absorption coefficient around a homogeneous optical property value, or similarly the distribution of the fluorochromes inside the medium. For fluorescence, the fluence rate at one point in the medium $\vec{r}'$ produced by the perturbation $O(\vec{r}')$ can be written in terms of the incident field $\phi_0(\vec{r}')$ as [24, 37]

$$\phi_{sc}(\vec{r}') = O^f(\vec{r}')\phi(\vec{r}', \vec{r}_s, \omega). \tag{12.14}$$

The measured field at the boundary of the scattering medium can be derived by integrating Equation 12.14 over the complete medium and propagating the fluence rate from the fluorochrome to the boundary:

$$\phi_{sc}(\vec{r}, \vec{r}_s, \omega) = \int g(\vec{r}, \vec{r}', \omega) O^f(\vec{r}')\phi(\vec{r}', \vec{r}_s, \omega) d\vec{r}' \tag{12.15}$$

where $g(\vec{r}, \vec{r}_s, \omega)$ is the Green's function solution of the diffusion equation for a single delta function at $\vec{r}'$. To solve Equation 12.15 bears the difficulty that the field $\phi(\vec{r}, \vec{r}')$ is generally a function of $\phi_{sc}(\vec{r}, \vec{r}_s)$, because this photon distribution depends on $O(r)$. To solve Equation 12.15 analytically, a linearization is performed using an approximation such as the Born or the Rytov approximation [24, 36]. In both approximations, the quantity $\phi(\vec{r})$ is essentially assumed equal to the photon field $\phi_0(\vec{r})$ that is established by the source at r_s in a similar medium without the heterogeneity present. As derived from the Born approximation, Equation 12.15 can then be written as

$$\phi_{sc}(\vec{r}, \vec{r}_s, \omega) = \int g(\vec{r}, \vec{r}', \omega) O^f(\vec{r}')\phi_0(\vec{r}', \vec{r}_s, \omega) d\vec{r}'. \tag{12.16}$$

The difference between this equation and Equation 12.15 is the replacement of $\phi(\vec{r}, \vec{r}', \omega)d\vec{r}'$ by $\phi_0(\vec{r}, \vec{r}', \omega)d\vec{r}'$. We can further refine this expression when considering that the light traveling toward the fluorochrome has the excitation wavelength λ_{ex}, and the light traveling from the fluorochrome to the boundary travels at the emission wavelength of the fluorochrome λ_{em} [9, 37]:

$$\phi_{fl}(\vec{r}, \vec{r}_s, \omega) = \int g^{\lambda_{em}}(\vec{r}', \vec{r}_s, \omega) O^f(\vec{r}', \omega)\phi_0^{\lambda_{ex}}(\vec{r}, \vec{r}', \omega) d\vec{r}'. \tag{12.17}$$

The distribution of fluorescence O^f is complex and depends on the modulation frequency of the source as well as on the quantum yield γ of the fluorochrome, the extinction coefficient of the fluorochrome ϵ, the fluorochrome concentration $F(r)$ and the fluorochrome lifetime τ [31]:

$$O^f(\vec{r}) \sim \frac{\gamma\epsilon[F(r')]}{1 - i\omega\tau}. \tag{12.18}$$

12.2.2 Reconstruction of the fluorochrome distribution

Fluorescence measurements can always be referenced to excitation measurements obtained under identical experimental characteristics by use of different filter sets; see Figure 12.5. Advantage of this additional information can be taken by normalizing the fluorescence measurement with the emission

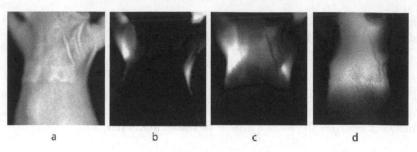

FIGURE 12.5: Measurements of a mouse with fluorescent biodistribution in the lung. (a) white light image; (b) transillumination excitation light measurement; (c) transillumination fluorescence measurement; (d) normalized fluorescence measurement.

measurements. One such approach is termed as the normalized Born approximation [35]. In this method, Equation 12.22 is divided by a measurement at the emission wavelength:

$$\frac{\phi_{em}(\vec{r}, \vec{r}_s, \omega)}{\phi_{ex}(\vec{r}, \vec{r}_s, \omega)} = \frac{\Theta}{\phi_0^{\lambda_{ex}}(\vec{r}', \vec{r}_s, \omega)} \int g^{\lambda_{em}}(\vec{r}, \vec{r}', \omega) O^f(\vec{r}', \omega) \phi_0^{\lambda_{ex}}(\vec{r}', \vec{r}_s, \omega) d\vec{r}'$$

(12.19)

where Θ is a constant that accounts for gain factors. This normalization eliminates instrumentation-related effects, and reduces the sensitivity of the reconstruction to errors in optical properties. For one source-detector pair the resulting linear problem formulated in terms of Green's functions is given by

$$\frac{\phi_{em}(\vec{r}_d, \vec{r}_s, \omega)}{\phi_{ex}(\vec{r}_d, \vec{r}_s, \omega)} = \sum \frac{G(\vec{r}', \vec{r}_s, \omega) n(\vec{r}') G(\vec{r}, \vec{r}_d, \omega)}{G(\vec{r}_d, \vec{r}_s, \omega)} dV.$$

(12.20)

The right-hand side is a sum over the voxels into which the imaged volume V is discretized. The Green's functions G can be computed using analytical methods [35, 37, 9] or numerical methods [23, 43] such as the Finite Element Method or Finite Volume Method. $G(\vec{r}', \vec{r}_s, \omega)$ represents the Green's function describing light propagating from source position r_s to position r inside the volume, $G(\vec{r}', \vec{r}_s, \omega)$ describes the light propagating from the point inside the volume to the detector position r_d and $G(\vec{r}_d, \vec{r}_s, \omega)$ is the normalization term. The volume of the voxels is included by the term dV and $n(\vec{r}')$ is the unknown fluorochrome distribution inside the volume. For the total number of source-detector pairs N_{data}, the resulting linear problem is written as

$$y = Wn$$

(12.21)

where y of size $1 \times N_{data}$ is the normalized data computed from the measurements at the surface, W of size $N_{data} \times N_{voxels}$ is called the weight matrix and n of size $1 \times N_{voxels}$ denotes the fluorescent source distribution.

An estimation of the fluorochrome distribution can be found by least squares minimization of the difference between the measured photon density y and the photon density as predicted by the forward model given by $\hat{y} = W\hat{x}$ with $\hat{x}$ the estimated fluorochrome concentration. The resulting function Q to be minimized is given by:

$$Q = \|Wx - y\|^2 + \lambda \|Lx\|^2. \tag{12.22}$$

Many different methods for solving this minimization problem exist [17]. An often-used approach is to set the regularization matrix L to the identity matrix I, resulting in a method called standard Tikhonov regularization. In case structural prior information is available, this can be included in the inverse problem by shaping the regularization matrix L in such a way that it reflects the structures that are available [15].

12.3 FMT and hybrid FMT systems

The basic elements of an FMT system are an illumination source, typically a laser with a wavelength in the near-infrared range, and a detector to record the light transmitted through the imaging subject. In the initial optical tomography systems, light was guided through fibers for illumination and detection [33]. The imaging subject would be submersed in fluids with the same absorption and scattering properties to improve fiber coupling, and to simplify the boundary conditions by submersing it in a case with a geometrically simple shape. Subsequently advances were made toward free-space technology [42, 45] to enable greater flexibility. The detection fibers were replaced by CCD cameras, and the illumination contact fibers by flying spot illumination; see Figure 12.6. The improvement of fluorescent targets for imaging purposes complements these developments in system technology. In the following we describe the current technology used in more detail.

12.3.1 Instrumentation

12.3.1.1 Illumination

An often-used method for illumination of tissue is to employ a source of constant intensity, generally termed CW, where CW stands for continuous wave. With this method it is possible to resolve changes in light attenuation. The main advantages of using constant wave technology for fluorescence biodistribution studies is that it offers good signal-to-noise characteristics and is generally operationally simple, low cost and robust [33]. The illumination source used for this domain is most often a diode laser source. Disadvantages are that it is difficult to resolve scattering and absorption characteristics using

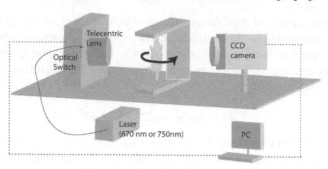

FIGURE 12.6: Setup for complete angle free-space tomography, based on a laser beam scanner and CCD camera detection. (From [33].)

this method. Two other possible approaches are the use of light of modulated intensity, typically at frequencies of 100MHz–1GHz or the use of ultrafast photon pulses in the 100 fs–100 ps range. Correspondingly, the detection systems can either measure changes in light attenuation and phase at different frequencies or offer ultrafast detection of photon kinetics with resolutions of the order of tenths of picoseconds or better. These time-domain and frequency-domain methods can also independently resolve fluorescence strength and lifetime [31].

12.3.1.2   Detection

The new generation of systems developed for fluorescence tomography is based on non-contact measurements, in which source and detector do not come in physical contact with the tissue [31, 16, 39]. Free-space detection is performed using charge coupled device (CCD) cameras. In a CCD camera, electromagnetic radiation is transformed into an electrical voltage. It consists of photoactive regions, "pixels," where incoming photons are absorbed, and a shift register to transfer charges from the photoactive regions to a charge amplifier where the charge is converted to a voltage. The photoactive regions are gates, each consisting of a metal-oxide-semiconductor capacitor. Photons that are absorbed in the semiconductor capacitor create electron-hole pairs. The photoelectrons are then transfered in sequence from one capacitor to the next, and finally arrive at the read-out point where the corresponding voltage per pixel is obtained. Using this technique, a much larger and higher-quality dataset can be obtained compared to fiber-based systems. It enables detection of signals of arbitrary shape and placement.

12.3.1.3   360° projections

In order to achieve superior imaging performance, the imaging subject needs to be illuminated from a large number of angles, similar to other tomographic techniques such as X-ray CT, PET, or SPECT. Several setups can be thought of to obtain 360° full angular coverage projections; the imaging subject can be rotated while placed in the middle of the instrumentation, or in a more advanced approach the instrumentation can be rotated instead. In case of rotating instrumentation, the animal is assumed to be in a relatively more comfortable position, which increases the possibilities of *in vivo* experiments, with greater flexibility to anesthetize the animal, and the possibility to monitor disease progression in an animal over time by repeated imaging. This last approach is attractive for multimodal imaging solutions.

12.3.2 Multimodal optical imaging

The latest developments in the field of optical imaging include the development of multi-modality systems. Before the development of hybrid systems, co-registration was achieved for example by using appropriate mouse constrainers to transfer mice from one modality to another without changing their placement or shape. But because FMT is based on the use of safe, non-ionizing energies, it is compatible with many different imaging modalities, and can be integrated with other modalities in one hybrid system. The simplest form of co-registration is by optically capturing the tissue outline or surface. Knowledge of the boundary of the imaging volume is a great advantage in free-space imaging of arbitrary shapes, as it can be used to more accurately calculate the forward model of light propagation which is used for reconstruction of the fluorescent source distribution. Combination of fluorescence tomography modalities with three-dimensional imaging modalities such as MRI or X-ray CT will yield even more benefit, since both modalities have complementing features. Fluorescence tomography offers depth-dependent fluorescence contrast which is mainly functional or molecular, but it does not give any anatomical information. Therefore co-registration with modalities that reveal anatomy will help in positioning and understanding the source of contrast. The images of the fluorescent source distribution can be superimposed for improved visualization. Furthermore, the anatomical modality can be used to guide the inversion problem by offering *a priori* information to facilitate the reconstruction of the fluorochrome distribution solution.

12.3.2.1 Optical tomography and MRI

The combination of near-infrared optical tomography and MRI has been explored in the context of breast cancer detection and pre-clinical research [6, 29, 30]. In breast cancer detection applications, imaging is focused on the reconstruction of the optical property distribution of the breast, specifically of absorption and scattering maps. Different optical coefficients can be coupled to different tissue types (tumor, background, etc.). The MRI image can subsequently be used to create a segmentation of the structures indicated by the optical property map in order to identify malignant regions. The MRI image can also be used other way around, to guide the reconstruction of the optical property distribution through regularization. In that case the MRI image is

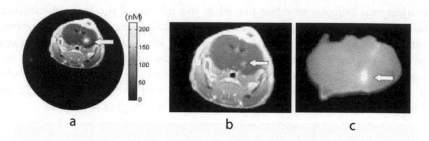

a   b   c

FIGURE 12.7: (See color insert.) Combined FMT and MRI imaging. Examples of fluorescence imaging using a cathepsin B-activatable imaging probe. (a) and (b), Enzyme activity in a 9L glioma model in a live mouse. The image in (a) is superimposed onto an MRI image shown separately in (b) with gadolinium enhancement of the glioma28. (c), In vitro FRI of the axial brain section corresponding to the MR and FMT images. The tumor position is indicated by the arrow. (From [29].)

segmented in different regions first, and the segment boundaries are used as prior information for calculation of the optical property map. An example of pre-clinical combined FMT and MRI imaging is illustrated in Figure 12.7. In this experiment the FMT image of a brain tumor was superimposed on the corresponding MRI image. The new generation of hybrid optical tomography and MRI systems in which the instrumentation is fully integrated [56] consist of an MRI scanner with laser diodes at one or more wavelengths for optical illumination guided into the MRI coil using fibers.

12.3.2.2 FMT-XCT

A very promising hybrid approach is the combination of free-space fluorescence tomography instrumentation with X-ray CT. Figures 12.8 and 12.9 show results of an experiment performed with a nude mouse with lung tumors. FMT and XCT data were coregistered and later combined in this case. Due to the similar configuration characteristics, these modalities are ideal for combination into one single system.

An example of a hybrid FMT-XCT system is schematically represented in Figure 12.10 [46]. In this setup, free-space FMT instrumentation is mounted on the same rotating gantry as the X-ray CT components. The X-ray components consist of an X-ray source and an X-ray detector. The FMT instrumentation includes a CCD camera coupled to a macro lens and diode laser sources, placed orthogonal to the X-ray source-detector axis. The FMT instrumentation needs to be protected from radiation using shielding. In front of the CCD camera lens a filter wheel is positioned, which includes a lead filter for radiation shielding during X-ray exposure, and different combinations of fil-

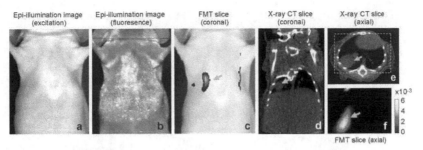

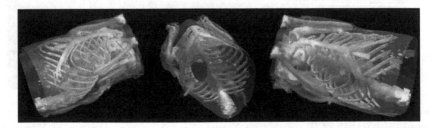

Epi-illumination image (excitation) — Epi-illumination image (fluoresence) — FMT slice (coronal) — X-ray CT slice (coronal) — X-ray CT slice (axial)

a   b   c   d   e

×10⁻³
6
4
2
0

f

FMT slice (axial)

FIGURE 12.8: (See color insert.) Tomographic imaging of fluorescent proteins and corresponding X-ray CT from a nude mouse implanted with GFP-expressing lung tumors, obtained 10 days post-image implantation. (a) Epi-illumination image of the mouse at the excitation wavelength; (b) Epi-illumination image at the emission wavelength showing high skin autofluorescence. (c) Tomographic slice (in color, after threshold was applied) obtained from the tumor depth (7 mm from top surface) overlaid on the white light image of the mouse. (d, c) CT coronal and axial slices, respectively; the tumor position is marked by arrows. (f) Axially reconstructed slice corresponding to the yellow dashed rectangle on (e). (From [31].)

ters to filter for the excitation or fluorescence wavelength. This type of hybrid FMT-XCT scanners allows 360-degree projection viewing for both FMT and XCT and the use of CCD cameras for photon detection leads to high spatial sampling of photon fields propagating through tissues such as small animals. Combination of the FMT and XCT modality into a single system eliminates the need to transfer the imaging subject from one system to the other using a specially designed case or markers for co-localization; the imaging subject is optimally co-localized in time and space, while it is placed on a comfortable bed. Reconstruction algorithms that are being developed specifically for datasets from this type of system use the X-ray data to improve FMT imaging quality in several aspects: the anatomy is used for assignment of optical

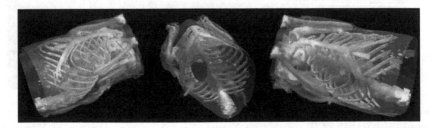

FIGURE 12.9: (See color insert.) Multimodality imaging. The fluorescent reconstructions of Figure 12.8 rendered simultaneously with X-ray CT images. The tumor is indicated by an arrow. (From [31].)

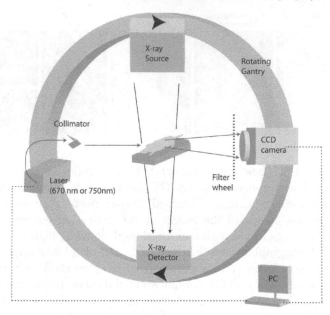

FIGURE 12.10: Combined FMT-XCT setup.

properties, structural priors based on the anatomy are used to guide the reconstruction and the XCT images are used for combined 3D visualization of the fluorescence signal together with anatomy [1, 22]. Figure 12.11 shows results obtained with a hybrid FMT-XCT system. The anatomical segmentation

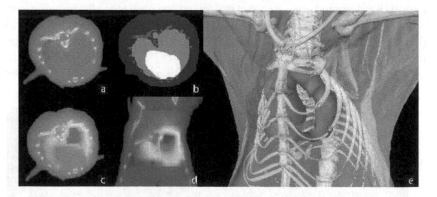

FIGURE 12.11: (See color insert.) Reconstruction based on data from combined FMT-XCT setup. (a) X-ray slice. (b) Segmentation of X-ray data in lungs, heart, bone and remaining tissue. (c) Reconstruction of fluorescent biodistribution in lung, transversal slice and (d) sagittal slice. (e) 3D-hybrid visualization. (From [22].)

based on the X-ray data is used as prior information in the reconstruction of the fluorescence biodistribution as described above. Accurate reconstructions of the fluorescence biodistribution in combination with hybrid 3D visualization can greatly improve our understanding of molecular processes *in vivo*.

References

[1] A. Ale, R. B. Schulz, A. Sarantopoulos, and V. Ntziachristos. Imaging performance of a hybrid x-ray computed tomography-fluorescence molecular tomography system using priors. *Medical Imaging*, 37(5):1976–1986, 2010.

[2] G. Alexandrakis, E. B. Brown, R. T. Tong, T. D. McKee, R. B. Campbell, Y. Boucher, and R. K. Jain. Two-photon fluorescence correlation microscopy reveals the two-phase nature of transport in tumors. *Nature Medicine*, 10(2):203–207, 2004.

[3] S. R. Arridge. Optical tomography in medical imaging. *Inverse Problems*, 15:41–93, 1999.

[4] D. A. Boas, D. H. Brooks, E. L. Miller, C. A. DiMarzio, M. Kilmer, R. J. Gaudette, and Q. Zhang. Imaging the body with diffuse optical tomography. *IEEE Signal Processing Magazine*, 18(6):57–75, 2001.

[5] C. F. Bohren and D. R. Huffman. *Absorption and Scattering of Light by Small Particles*. John Wiley & Sons, Inc., New York, 1983.

[6] B. A. Brooksby, H. Dehgani, B. W. Pogue, and K. D. Paulsen. Near-infrared (NIR) tomography breast image reconstruction with a priori structural information from MRI: Algorithm development for reconstructing heterogeneities. *IEEE Journal of Selected Topics Quantum Electronics*, 9(2):199–209, 2003.

[7] K. M. Case and P. F. Zweifel. *Linear Transport Theory*. Addison-Wesley Series in Nuclear Engineering. Addison-Wesley, Reading, MA, 1967.

[8] S. Chandraeskhar. *Radiative Transfer*. Dover Publications, New York, 1960.

[9] J. W. Chang, H. L. Graber, and R. L. Barbour. Luminescence optical tomography of dense scattering media. *Journal of the Optical Society of America A-Optics Image Science and Vision*, 14(1):288–299, 1997.

[10] X. Y. Chen, P. S. Conti, and R. A. Moats. In vivo near-infrared fluorescence imaging of integrin $\alpha_v\beta_3$ in brain tumor xenografts. *Cancer Research*, 64(21):8009–8014, 2004.

[11] B. Davison. *Neutron Transport Theory.* International Series of Monographs on Physics. Clarendon Press, Oxford, UK, 1957.

[12] J. J. Duderstadt and W. R. Martin. *Transport Theory.* John Wiley & Sons, Inc., New York, 1979.

[13] M. Fox. *Optical Properties of Solids.* Oxford University Press, New York, 2001.

[14] A. Gibson and H. Dehghani. Diffuse optical imaging. *Philosophical Transactions of the Royal Society A—Mathematical Physical and Engineering Sciences,* 367(1900):3055–3072, 2009.

[15] A. Gibson, J. C. Hebde, and S. R. Arridge. Recent advances in diffuse optical imaging. *Physics in Medicine and Biology,* 50:1–43, 2005.

[16] E. Graves, J. Ripoll, R. Weissleder, and V. Ntziachristos. A submilimeter resolution fluorescence molecular imaging system for small animal imaging. *Medical Physics,* 30:901–911, 2003.

[17] P. C. Hansen. *Rank-Deficient and Discrete Ill-Posed Problems.* SIAM, Philadelphia, 1998.

[18] E. Hecht. *Optics.* Addison-Wesley, Reading, MA, 2002.

[19] S. W. Hell. Microscopy and its focal switch. *Nature Methods,* 6(1):24–32, 2009.

[20] B. Huang, M. Bates, and X. W. Zhuang. Super-resolution fluorescence microscopy. *Annual Review of Biochemistry,* 78:993–1016, 2009.

[21] J. Huisken, J. Swoger, F. Del Bene, J. Wittbrodt, and E. H. K. Stelzer. Optical sectioning deep inside live embryos by selective plane illumination microscopy. *Science,* 305(5686):1007–1009, 2004.

[22] D. Hyde, E. L. Miller, D. H. Brooks, and V. Ntziachristos. Data specific spatially varying regularization for multimodal fluorescence molecular tomography. *IEEE Transactions on Medical Imaging,* 29(2):365–374, 2010.

[23] A. Joshi, W. Bangerth, and E. M. Sevick-Muraca. Adaptive finite element based tomography for fluorescence optical imaging in tissue. *Optics Express,* 12(22):5402–5417, 2004.

[24] A. Kak and M. Slaney. *Principles of Computerized Tomographic Imaging.* IEEE Press, New York, 1988.

[25] T. Lin, H. Yan, O Nalcioglu, and G. Gulsen. Quantitative fluorescence tomography with functional and structural a priori information. *Applied Optics,* 48(7):1328–1336, 2009.

[26] R. Loudon. *The Quantum Theory of Light*. Oxford Science Publications. Clarendon Press, Oxford University Press, 1983.

[27] W. K. Moon, Y. H. Lin, T. O'Loughlin, Y. Tang, D. E. Kim, R. Weissleder, and C. H. Tung. Enhanced tumor detection using a folate receptor-targeted near-infrared fluorochrome conjugate. *Bioconjugate Chemistry*, 14(3):539–545, 2003.

[28] M. Niedre and V. Ntziachristos. Comparison of fluorescence tomographic imaging in mice with early-arriving and quasi-continuous-wave photons. *Optics Letters*, 35(3):369–371, 2010.

[29] B. Ntziachristos, C. Tung, C. Bremer, and R. Weissleder. Fluorescence-mediated tomography resolves protease activity in vivo. *Nature Medicine*, 8(7):757–760, 2002.

[30] B. Ntziachristos, A. G. Yodh, M. Schnall, and B. Chance. Concurrent MRI and diffuse optical tomography of breast after indocyanine green enhancement. *Proceedings of the National Academy of Science of the United States of America*, 97(6):2767–2772, 2000.

[31] V. Ntziachristos. Fluorescence molecular imaging. *Annual Review of Biomedical Engineering*, 8:1–33, 2006.

[32] V. Ntziachristos and B. Chance. Probing physiology and molecular function using optical imaging: applications to breast cancer. *Breast Cancer Research*, 3(1):41–46, 2001.

[33] V. Ntziachristos, A. Leroy-Willig, and B. Tavitian, editors. *Textbook of in vivo Imaging in Vertebrates*. John Wiley & Sons Ltd., New York, 2007.

[34] V. Ntziachristos, J. Ripoll, L. H. V. Wang, and R. Weissleder. Looking and listening to light: The evolution of whole-body photonic imaging. *Nature Biotechnology*, 23(3):313–320, 2005.

[35] V. Ntziachristos and R. Weissleder. Experimental three-dimensional fluorescence reconstruction of diffuse media by use of a normalized Born approximation. *Optics Letters*, 26(12):893–895, 2001.

[36] M. A. O'Leary, D. A. Boas, B. Chance, and A. G. Yodh. Experimental images of heterogeneous turbid media by frequency-domain diffusing-photon tomography. *Optics Letters*, 20:426–428, 1995.

[37] M. A. O'Leary, D. A. Boas, X. D. Li, B. Chance, and A. G. Yodh. Fluorescence lifetime imaging in turbid media. *Optics Letters*, 21:158–160, 1996.

[38] M. S. Patterson, B. Chance, and B. C. Wilson. Time-resolved reflectance and transmittance for the noninvasive measurement of tissue optical properties. *Applied Optics*, 28(12):2331–2336, 1989.

[39] S. V. Patwardhan, S. R. Bloch, S. Achilefu, and J. P. Culver. Time-dependent whole-body fluorescence tomography of probe bio-distributions in mice. *Optics Express*, 13:2546–2577, 2005.

[40] D. Piwnica-Worms, D. P. Schuster, and J. R. Garbow. Molecular imaging of host-pathogen interactions in intact small animals. *Cellular Microbiology*, 6(4):319–331, 2004.

[41] D. Razansky, C. Vinegoni, and V. Ntziachristos. Imaging of mesoscopic-scale organisms using selective-plane optoacoustic tomography. *Physics in Medicine and Biology*, 54(9):2769–2777, 2009.

[42] J. Ripoll, R. B. Schulz, and V. Ntziachristos. Free-space propagation of diffuse light: Theory and experiments. *Physical Review Letters*, 91(10):103901–103904, 2003.

[43] R. Roy and E. M. Sevick-Muraca. Three-dimensional unconstrained and constrained image-reconstruction techniques applied to fluorescence, frequency-domain photon migration. *Applied Optics*, 40(13):2206–2215, 2001.

[44] S. M. Rytov, Y. A. Kravtsov, and V. I. Tararskii. *Principles of Statistical Radiophysics*. Springer-Verlag, Berlin, 1987.

[45] R. B. Schulz, J. Ripoll, and V. Ntziachristos. Experimental fluorescence tomography of tissues with noncontact measurements. *IEEE Transactions on Medical Imaging*, 23(4):492–500, 2004.

[46] R. B. Schulz, A. Ale, A. Sarantopoulos, M. Freyer, E. Soehngen, M. Zientkowska, and V. Ntziachristos. Hybrid system for simultaneous fluorescence and x-ray computed tomography. *IEEE Transactions on Medical Imaging*, 29(2):456–473, 2010.

[47] J. Sharpe, U. Ahlgren, P. Perry, B. Hill, A. Ross, J. Hecksher-Sorensen, R. Baldock, and D. Davidson. Optical projection tomography as a tool for 3D microscopy and gene expression studies. *Science*, 296(5567):541–545, 2002.

[48] P. Sheng. *Scattering and Localization of Classical Waves in Random Media*. Volume 8 of World Scientific Series on Directions in Condensed Matter Physics. World Scientific, Singapore, 1990.

[49] V.V. Sobolev. *Light Scattering in Planetary Atmospheres*. Volume 76 of International Series of Monographs in Natural Philosophy. Pergamon Press, Oxford, UK, 1975.

[50] D. J. Stephens and V. J. Allan. Light microscopy techniques for live cell imaging. *Science*, 300(5616):82–86, 2003.

[51] D. Toomre and D. J. Manstein. Lighting up the cell surface with evanescent wave microscopy. *Trends in Cell Biology*, 11(7):298–303, 2001.

[52] L. Tsang and J. A. Kong. *Scattering of Electromagnetic Waves*. Wiley Series in Remote Sensing. John Wiley & Sons, Inc., New York, 2001.

[53] L. Tsang, J. A. Kong, and K.-H. Ding. *Scattering of Electromagnetic Waves*. Wiley Series in Remote Sensing. John Wiley & Sons, Inc., New York, 2000.

[54] L. Tsang, J. A. Kong, and R. T. Shin. *Theory of Microwave Remote Sensing*. Wiley Series in Remote Sensing. John Wiley & Sons, Inc., New York, 1985.

[55] B. Valuer. *Molecular Fluorescence: Principles and Applications*. Wiley-VCH, Weinheim, Germany, 2002.

[56] J. Wang, B. W. Pogue, S. Jiang, and K. D. Paulsen. Near-infrared tomography of breast cancer hemoglobin, water, lipid, and scattering using combined frequency domain and CW measurement. *Optics Letters*, 35(1):82–84, 2010.

[57] L. V. Wang and H. Wu. *Biomedical Optics Principles and Imaging*. John Wiley & Sons, Inc., Hoboken, NJ, 2007.

[58] R Weissleder and V. Ntziachristos. Shedding light onto live molecular targers. *Nature Medicine*, 9(1):123–128, 2003.

[59] A. Wunder, C. H. Tung, U. Muller-Ladner, R. Weissleder, and U. Mahmood. In vivo imaging of protease activity in arthritis: A novel approach for monitoring treatment response. *Arthritis and Rheumatism*, 50(8):2459–2465, 2004.

[60] W. R. Zipfel, R. M. Williams, and W. W. Webb. Nonlinear magic: Multiphoton microscopy in the biosciences. *Nature Biotechnology*, 21(11):1368–1376, 2003.

[61] A. Zoumi, A. Yeh, and B. J. Tromberg. Imaging cells and extracellular matrix in vivo by using second-harmonic generation and two-photon excited fluorescence. *Proceedings of the National Academy of Sciences of the United States of America*, 99(17):11014–11019, 2002.

Index

CT, 167
LOR, 161
MRI, 167
PET, 167
2D rebinning, 39
3D filtered backprojection, 40

Additive noise, 121
Affine motion correction, 163
Algorithm
    Bruhn, 170
    discontinuity preserving, 172
    Horn/Schunck, 169
    Lucas/Kanade, 168
    mass conserving OF, 174
Analytical algorithm, 32
ART, 41, 44
Attenuation, 54, 73, 82

Backprojection, 33
Bioluminescence imaging, 18
Biomedical application, 14
Blood-brain barrier, 16
Brightness consistency constraint, 165

Catecholamine, 22
Clinical application, 17
Clinical decision making, 11
Collimator effects, 62
Compartment modeling, 187
Compton scattering, 52
Correction techniques, 4
Crystal efficiency, 108

Dead time, 59, 108
Decay correction, 69
Deconvolution, 124, 142
    blind deconvolution, 143

Richardson–Lucy deconvolution,
    142
    Van-Cittert deconvolution, 142
Denoising, 121
Depth-of-interaction, 106
Discontinuity preserving, 172
Discrete Algorithms, 40

Elastic motion correction, 163
Electromagnetic spectrum, 2, 4
EM, 42
EM List Mode, 46

Filtered backprojection, 35
Filtering
    Gaussian filtering, 122
    mean filtering, 122
    median filtering, 122
    Wiener filtering, 122
Flow field, 164
Fourier transform, 34, 35, 123
Full width at half maximum (FWHM),
    138

Gating, 161
Graft-versus-host disease, 20

Horn/Schunck, 169
Hybrid imaging, 209

Image constraint equation, 164
Image reconstruction, 31
Image registration, 120
Image restoration, 121
Interpolation, 120, 126
    linear, 126
    nearest neighbor, 126
    sinc, 126

spline, 126

Kaczmarz algorithm, 41
Klein–Nishina, 53

Light output, 107
Line of response, 32
List mode, 45
Listmode, 193
Lucas–Kanade, 168

Magnetic resonance, 211
Mass conserving OF, 174
Maximum ring difference, 106
Mean-squared error, 123
Molecular imaging, 13
Motion correction, 114, 173, 194
    affine, 163
    elastic, 163
    reconstruction, 192
    rigid, 162
MR/PET, 212

Noise equivalent counts, 108
Noise removal, 120
Non-paralyzable, 108
Normalization, 58, 107, 108

Optical flow, 164
    Bruhn, 170
    discontinuity preserving, 172
    Horn/Schunck, 169
    Lucas/Kanade, 168
    mass conserving, 174
    methods, 166
OSEM, 44

Paralyzable, 109
Partial volume, 110
Partial volume correction, 137
Partial volume effect, 59, 120, 137, 160, 174
PET/CT, 209, 211
PET/MR, 212
Phantom, 148
    hardware, 148

software, 149
Pharmakodynamic studies, 19
Phenotyping, 16
Photoelectric effect, 52
Point spread function, 121
Poisson, 43
Positron range, 63

Radon, 32, 33
Randoms, 61, 71
Receptor, 22
Registration, 129
    non-parametric registration, 133
    parametric registration, 132
    software, 137
Regularization, 133
Rigid motion correction, 162

Scatter, 57
Signal-to-noise ratio, 121
Similarity measure, 133
    conditional mutual information, 135
    cross-cumulative residual entropy, 135
    mutual information, 134
    normalized cross-correlation, 134
    normalized gradient fields, 134
    sum of squared differences, 134
Single Slice Rebinning, 39
Span, 106
SPECT/CT, 211
Spill in, 60, 111
Spill-over, 139
Spillout, 111
Standardized uptake value, 138
Super-resolution, 144
System Matrix, 44

Tissue fraction, 139
Tomography, 3
Tracer principle, 15
Tumor, 23

US, 167

Wavelet transform, 124

X-ray transform, 33
X-rays, 12

Milton Keynes UK
Ingram Content Group UK Ltd.
UKHW040447071024
449327UK00020B/1067

9 780367 381448